T0331668

# ELEMENTARY TOPOLOGY AND APPLICATIONS

### Second Edition

# ELEMENTARY TOPOLOGY AND APPLICATIONS

## Second Edition

### Carlos R Borges

*University of California, Davis, USA*

 **World Scientific**

NEW JERSEY · LONDON · SINGAPORE · BEIJING · SHANGHAI · HONG KONG · TAIPEI · CHENNAI · TOKYO

*Published by*

World Scientific Publishing Co. Pte. Ltd.

5 Toh Tuck Link, Singapore 596224

*USA office:* 27 Warren Street, Suite 401-402, Hackensack, NJ 07601

*UK office:* 57 Shelton Street, Covent Garden, London WC2H 9HE

**Library of Congress Cataloging-in-Publication Data**

Names: Borges, Carlos R., 1939–    author.

Title: Elementary topology and applications / Carlos R. Borges, University of California, Davis, USA.

Description: Second edition. | New Jersey : World Scientific, [2021] |
    Includes bibliographical references and index.

Identifiers: LCCN 2021026278 (print) | LCCN 2021026279 (ebook) | ISBN 9789811237423 (hardcover) |
    ISBN 9789811237430 (ebook) | ISBN 9789811237447 (mobi)

Subjects: LCSH: Topology.

Classification: LCC QA611 .B652 2021 (print) | LCC QA611 (ebook) | DDC 514--dc23

LC record available at https://lccn.loc.gov/2021026278

LC ebook record available at https://lccn.loc.gov/2021026279

**British Library Cataloguing-in-Publication Data**

A catalogue record for this book is available from the British Library.

For any available supplementary material, please visit
https://www.worldscientific.com/worldscibooks/10.1142/12293#t=suppl

Printed in Singapore

Dedicated to
my parents, my son Carlos
and my daughter Mary Lou

# Contents

# Preface

I view this book as an introduction to Topology with major emphasis on applications; it will adequately prepare students for further work in many mathematical disciplines.

The material is organized so that one gets to significant applications quickly, with emphasis on the geometric understanding and use of new concepts. The theme of this book is that Topology really is the universal language of modern mathematics.

We assume that the reader has a good grasp of the fundamentals of Logic and Set Theory, even though a rather succinct review appears in the introductory Chapter 0. The reader should also be very familiar with elementary analysis. Some familiarity with Group Theory is required for Chapters 5, 6 and 8.

The problems which appear at the end of each chapter not only provide ample opportunity for applying the concepts and techniques just learned but also are used to introduce additional concepts and techniques which complement the text and point to further study elsewhere.

I am indebted to many students who, over many years, have made very useful suggestions on the presentation that follows. I am also indebted to Ida Orahood (who typed these chapters as needed for my classes). Lori Carranza and Josh Walters (from the University Typing Pool, who retyped Ida's work) who typed the first drafts of this book, and I am especially indebted to John Gehrmann who did the final revision of the text. I am also very indebted to Dr. Sunil Nair who encouraged me and helped me publish this book.

The first edition contains many gaffes. The truth be told, most are due to unusual circumstances: The chair of the Math Dept insisted that math typists could work only in Research papers. Finally, an excellent math typist was allowed to do the final version, and he corrected hundreds of gaffes (unfortunately, many were not saved!)

Carlos R. Borges
December 2020

## Second Edition Acknowledgments

My heartfelt thanks goes to Senior Editor Lakshmi Narayanan and the Senior Type-setter Janice Sim for their helpfulness and expertise, which are truly admirable, in producing this latest edition.

# Chapter 0

# Sets and Numbers

**Set Theory**

The main purpose of this section is to collect those precious gems of Set Theory—Relations, Functions and Inverse Functions—which will frequently be used throughout the text. The few extra comments are designed, either to bring out a convenient viewpoint, or to point out some pitfalls.

We prove none of the set-theoretical formulas that we mention, inasmuch that, whoever is ready for what lies ahead, can prove these quite easily.

## 0.1 Rudiments of Logic

In order that we may appreciate the subtleties of mathematical thought, we need to have at our disposal uniform and precise rules of mathematical reasoning. For example, the implication

If John weighs less than Joe and Joe weighs less
than John, then John and Joe have the same weight

might provoke some to argue that it is nonsense, because it is impossible that

John weighs less than Joe and Joe weighs less than John,

while others might argue that it is exactly this impossibility that makes the implication valid, inasmuch that what is intended by

If $A$ then $B$

is simply that

$A$ is false or $B$ is true.

This forces us to establish a universal understanding of our basic means of communication: We assume that all expressions that we consider are *true* or *false*, but not both.

1

**Variables.** We designate arbitrary expressions by capital letters of the English alphabet, with subscripts, if necessary:

$$A, B, C, \ldots, A_1, B_1, C_1, \ldots, A_2, B_2, C_2, \ldots$$

**Connectives.** We use the following connectives to form new expressions from any two expressions at our disposal. (For convenience, we let $\equiv$ stand for *means that* or *is the same* as):

$$\text{or} \equiv \vee$$

$$\text{and} \equiv \wedge$$

$$\text{if} \ldots \text{then} \ldots \equiv \ldots \text{implies} \ldots \equiv \Rightarrow$$

$$\text{if and only if} \equiv \text{iff} \equiv \Leftrightarrow$$

$$\sim \equiv \text{negation of} \equiv \text{not} -$$

**Truth Values.** We use only two truth values (More elaborate logical systems see the need for at least one more value—indeterminate.):

$$T \equiv \text{true}$$

$$F \equiv \text{false}$$

**Quantifiers.** The following symbols specify quantity:

$$\forall \equiv \text{for every} \equiv \text{for each}$$

$$\exists \equiv \text{for some} \equiv \text{there exists}$$

$$= \; \equiv \text{is equal to}$$

**Primitive Symbols.** These symbols help us avoid confusion in the communication of information:

$$( \; \equiv \text{left parenthesis}$$

$$) \equiv \text{right parenthesis}$$

**Truth Tables.** We now give the truth tables of the connectives. In this manner, we specify exactly what is our understanding of the connectives $\vee, \wedge, \Rightarrow, \Leftrightarrow, \sim$ in mathematical reasoning; some of these do have vague and ambiguous interpretations in their quotidian use.

The truth tables of the following expressions will guide us into some important rules of logical reasoning:

| $A$ | $\sim A$ |
|---|---|
| T | F |
| F | T |

| $A$ | $B$ | $A \vee B$ |
|---|---|---|
| T | T | T |
| T | F | T |
| F | T | T |
| F | F | F |

| $A$ | $B$ | $A \wedge B$ |
|---|---|---|
| T | T | T |
| T | F | F |
| F | T | F |
| F | F | F |

| $A$ | $B$ | $A \Rightarrow B$ |
|---|---|---|
| T | T | T |
| T | F | F |
| F | T | T |
| F | F | T |

| $A$ | $B$ | $A \Leftrightarrow B$ |
|---|---|---|
| T | T | T |
| T | F | F |
| F | T | F |
| F | F | T |

| $A$ | $B$ | $\sim A$ | $\sim B$ | $A \Rightarrow B$ | $\sim B \Rightarrow \sim A$ | $(A \Rightarrow B) \Leftrightarrow (\sim B \Rightarrow \sim A)$ |
|---|---|---|---|---|---|---|
| T | T | F | F | T | T | T |
| T | F | F | T | F | F | T |
| F | T | T | F | T | T | T |
| F | F | T | T | T | T | T |

| $A$ | $B$ | $\sim A$ | $A \Rightarrow B$ | $\sim A \vee B$ | $(A \Rightarrow B) \Leftrightarrow (\sim A \vee B)$ |
|---|---|---|---|---|---|
| T | T | F | T | T | T |
| T | F | F | F | F | T |
| F | T | T | T | T | T |
| F | F | T | T | T | T |

| $A$ | $B$ | $\sim A$ | $\sim B$ | $A \vee B$ | $\sim (A \vee B)$ | $(\sim A) \wedge (\sim B)$ | $\sim (A \vee B) \Leftrightarrow (\sim A) \wedge (\sim B)$ |
|---|---|---|---|---|---|---|---|
| T | T | F | F | T | F | F | T |
| T | F | F | T | T | F | F | T |
| F | T | T | F | T | F | F | T |
| F | F | T | T | F | T | T | T |

| $A$ | $B$ | $\sim A$ | $\sim B$ | $A \wedge B$ | $\sim (A \wedge B)$ | $(\sim A) \vee (\sim B)$ | $\sim (A \wedge B) \Leftrightarrow (\sim A) \vee (\sim B)$ |
|---|---|---|---|---|---|---|---|
| T | T | F | F | T | F | F | T |
| T | F | F | T | F | T | T | T |
| F | T | T | F | F | T | T | T |
| F | F | T | T | F | T | T | T |

We therefore conclude that the expressions (i)–(iv) are always true regardless of the truth values attributed to their variables (such expressions are called *tautologies*).

i) $(A \Rightarrow B) \quad \Leftrightarrow (\sim B \Rightarrow \sim A)$

ii) $(A \Rightarrow B) \quad \Leftrightarrow (\sim A \vee B)$

iii) $\sim (A \vee B) \Leftrightarrow (\sim A) \wedge (\sim B)$

iv) $\sim (A \wedge B) \Leftrightarrow (\sim A) \vee (\sim B)$

In accordance with the truth table for $\Leftrightarrow$, this means that, for any truth values for $A$ and $B$ the corresponding truth values of

$$A \Rightarrow B \quad \text{and} \quad \sim B \Rightarrow \sim A$$
$$A \Rightarrow B \quad \text{and} \quad \sim A \vee B$$
$$(\sim A \vee B) \quad \text{and} \quad (\sim A) \wedge (\sim B)$$
$$\sim (A \wedge B) \quad \text{and} \quad (\sim A) \vee (\sim B)$$

are, respectively, the same. Thus (ii) justifies the truth table we have adopted for $\Rightarrow$, since, when we say that "$A \Rightarrow B$ is true" we are really interested that $B$ be true or $A$ be false. Also, (i) justifies the *method of proof by the contrapositive* by which we prove that $A \Rightarrow B$ is true by showing that $\sim B \Rightarrow \sim A$ is true, instead.

Next, we establish the truth table of $A \wedge (\sim A)$ in order to explain another method of proof.

| $A$ | $\sim A$ | $A \wedge (\sim A)$ |
|:---:|:---:|:---:|
| T | F | F |
| F | T | F |

We see that, whatever truth values are given to $A$, $A \wedge (\sim A)$ is always false (such expressions are called *contradictions*).

We can justify the *method of proof by contradiction* by which we prove that an expression $A$ is true by exhibiting an expression $B$ such that

$$\sim A \Rightarrow (B \wedge (\sim B))$$

is true. Given that $B \wedge (\sim B)$ is false, we get from the truth table for $\Rightarrow$, that $\sim A$ is false; therefore, $A$ is true, because of the truth table for $\sim$.

We have thus completed our task of presenting the rudiments of a universal language that precisely states the elementary rules of mathematical reasoning. With these rules and an axiom system that tells us how to derive conclusions from given information, we could then proceed to the study of *logical reasoning*.

## 0.2 Fundamentals of Set Description

The concept of *set* is undefined; it is simply taken for granted that all (?) human beings, through the experiences of their daily lives, become aware of *sets* or *collections* or *groups* of objects—certainly, at one time or another we have all become painfully aware of crowd (large sets or groups of people). Do not be shocked by this; remember that, in Euclidean geometry, the concept of *line* and *point* are undefined.

One specifies a set by specifying its elements. A standard notation is

$$\{x|S(x)\} \equiv \{x : S(x)\}$$

which is read *the set of all elements $x$ such that $x$ satisfies the sentence $S(x)$*. Examples of the sentence $S(x)$ might be: "$x$ is a dog", "$x$ is a real number greater than zero" or "$x$ is a blonde mathematician with green eyes".

The difficulties in the development and understanding of set theory come mostly from the simple fact that not all sentences that one utters make sense. Obviously, the sentence "$x$ is a blonde with black hair" does not make sense. But there are nonsensical sentences that, at first sight, may appear perfectly acceptable: Let $S(x)$ be the sentence $x \notin x$. Let $A = \{x|x \notin x\}$. Then $A \in A$ implies that $A \notin A$, a contradiction; $A \notin A$ implies that $A \in A$, a contradiction. Since we follow a logical system that roughly says that *something either is or is not, but not both*, we then must consider the sentence $x \notin x$ unacceptable. Equally, we cannot accept sentences such as

$$S(x) \equiv \text{for some } x, \ x \text{ is a dog},$$

$$S(x) \equiv \text{for all } x, \ x \text{ is not a real number},$$

inasmuch that, for example, *the set of all $x$ such that for some $x$, $x$ is a dog* is an ambiguous statement. Therefore, *an acceptable sentence $S(x)$ must contain the variable $x$, completely free of the quantifiers $\forall$ and $\exists$, at least once.*

## 0.3 Set Inclusion and Equality

Given the sets $A$ and $B$, we say that

(i) $A$ is contained in $B$, provided that $x \in A$ implies that $x \in B$.

We let $A$ is not contained in $B \equiv A \not\subset B$.

(ii) $A$ contains $B \equiv B$ is contained in $A \equiv A \supset B$.

(iii) $A$ *is identical to* $B$ provided that $x \in A$ iff $x \in B$; equivalently, $A \subset B$ and $B \subset A$.

(We let $A$ is identical to $B \equiv A = B$.)

## 0.4 An Axiom System for Set Theory

Presently, there are various axiom systems for set theory. Fortunately, their differences are rather minute. We like, what might be called, the Gödel-Bernays system of six axioms. We state the first five now, leaving the sixth to be stated after we realize that there is a need for it. (This helps to make it more self-evident.)

I. (*Axiom of Extension.*) If the sets $A$ and $B$ have the same elements then they are identical.

II. (*Axiom of the Empty Set.*) There exists a set $\emptyset$ with no elements.

III. (*Axiom of Unions.*) Let $a$ be a set whose elements are sets. There exists a set $S$ such that $x \in S$ iff $x \in A$ for some $A \in a$.

IV. (*Axiom of Power Sets.*) For every set $X$ there exists a set $\mathfrak{p}(X)$ which consists exactly of all the subsets of the set $X$. $\mathfrak{p}(X)$ is called the *power set* of $X$.

V. (*Axiom of Infinity.*) There exists a set $S$ satisfying the conditions: Each element of $S$ is a set; $\emptyset \in S$; $m \in S$ implies that there is $n \in S$ which has only $m$ and the elements of $m$.

The axiom of infinity allows one to define the *natural members* and to establish the *principle of induction*, as we shall soon see.

## 0.5   Unions and Intersections

Given a collection $\mathcal{C}$ of sets,

$$\bigcup \mathcal{C} = \{x | X \in X \text{ for some } X \in \mathcal{C}\},$$

$$\bigcap \mathcal{C} = \{x | x \in \bigcup \mathcal{C} \text{ and } C \in \mathcal{C} \Rightarrow X \in C\}.$$

The following notation is also commonly used:

$$\bigcup \mathcal{C} = \bigcup \{X | X \in \mathcal{C}\} = \bigcup_{X \in \mathcal{C}} X, \bigcap \mathcal{C} = \bigcap \{X | X \in \mathcal{C}\} = \bigcap_{X \in \mathcal{C}} X.$$

If $\mathcal{C} = \emptyset$ then $\bigcup \emptyset = \emptyset$, $\bigcap \emptyset = \emptyset$.

**Observation.** An apparently insignificant alteration in the definition of intersection that we have adopted, produces dramatic effects: Let us suppose that, for each collection $\mathcal{C}$ of sets, we define

$$\bigcap \mathcal{C} = \{x | C \in \mathcal{C} \Rightarrow x \in C\}.$$

It follows that, if $\mathcal{C} \neq \emptyset$ then

$$\bigcap \mathcal{C} = \{x | x \in \bigcup \mathcal{C}, C \in \mathcal{C} \Rightarrow x \in C\} = \{x | C \in \mathcal{C} \Rightarrow x \in C\};$$

that is, the two definitions of intersection are identical, whenever $\mathcal{C} \neq \emptyset$. However, if $\mathcal{C} = \emptyset$, then the implication $C \in \mathcal{C} \Rightarrow x \in C$ becomes the *true* implication $C \in \emptyset \Rightarrow x \in C$ (because "$C \in \emptyset$" is false). Consequently $\{x | C \in \mathcal{C} \Rightarrow x \in C\} = \{x | C \in \emptyset \Rightarrow x \in C\}$ is a set. This implies that $\{x | x \notin x\}$ is a set, by the Axiom of Power sets. (Clearly $\{x | x \notin x\}$ is a subset of $\{x | C \in \emptyset \Rightarrow x \in C\}$, since the implication $C \in \emptyset \Rightarrow x \in C$ is true for all $x$, *including those* $x$ such that $x \notin x$.) But we already know that $\{x | x \notin x\}$ cannot be a set. If $\mathcal{C} = \{A\}$ (*i.e.* the collection $\mathcal{C}$ consists of a single set) then

$$\bigcup \{A\} = \bigcup \mathcal{C} = A \text{ and } \bigcap \{A\} = \bigcap \mathcal{C} = A.$$

## 0.6 Set Difference

For any sets $A$ and $B$,

$$A - B = \{x \in A | x \notin B\} = \{x | x \in A \text{ and } x \notin B\}.$$

(The Axiom of Power Sets easily assures us that $A - B$ is indeed a set.)

## 0.7 Integers and Induction

For any set $S$, let $S^+ = S \bigcup \{S\}$ (the set $S^+$ is called the *successor* of $S$) and, for the sake of familiarity, let $\emptyset \equiv 0$, in this context. The axiom of infinity implies the following:

**1. Proposition.** There exists exactly one collection $\omega$ of sets such that

(i) $0 \in \omega$

(ii) $n \in \omega \Rightarrow n^+ \in \omega$

(iii) if $K$ satisfies (i) and (ii) then $\omega \subset K$. (The set $\omega$ is called the *set of natural numbers*.)

**Proof.** By the axiom of infinity, there exists a family $\mathcal{F}$ satisfying (i) and (ii). Let $\Psi = \{S | S \subset \mathcal{F} \text{ and } S \text{ satisfies (i) and (ii)}\}$. It is easy to see that $\omega = \bigcap \Psi$ (note that $\Psi \neq \emptyset$).

Note that part (iii) of the preceding Proposition is the *Principle of Induction*. Indeed it is customary to call a set $K$ an *inductive set* provided that $K$ satisfies (i) and (ii) above.

By induction, it is easy to prove that

(iv) $m \in n$ implies $m^+ \subset n$; $m \in n$ implies $m^+ \in n^+$

(Show $K = \{n | m \in n \text{ implies } m^+ \subset n\}$

and $K = \{n | m \in n \text{ implies } m^+ \in n^+\}$ are inductive.).

(v) $n \notin n$, for each $n \in \omega$.

(vi) For $m, n \in \omega$ one and only one of the following holds: $m \in n$ or $n \in m$ or $m = n$. (Show $K = \{m | n \in \omega \text{ implies one and only one of } m \in n \text{ or } n \in m \text{ or } m = n\}$ is an inductive set, by the use of (iv) and (v).)

(vii) $m^+ = n^+ \Rightarrow m = n$

(viii) $n^+ \neq 0$, for every $n \in \omega$

(Show $K = \{0\} \bigcup \{n | n^+ \neq 0\}$ is an inductive set.)

(ix) $n \in \omega$, $n \neq 0$ implies that there exists $k \in \omega$ such that $k^+ = n$. ($k$ is said to be the *predecessor* of $n$; it is immediate that $k$ is unique, by (vii); it is customary to denote the predecessor of $n$ by $n^-$ or $n - 1$).

(x) For each $m \in \omega$, $\{n \in \omega | n \in m\} = m$.
(Show $K = \{m \in \omega | m = \{n \in \omega | n \in m\}\}$ is inductive. This property of $m$ is very important and is related to the sixth axiom.)

(xi) If $u = \{n \in \omega | n \in u\}$ and there exists $t \in \omega$ such that $u \in t$, then $u \in \omega$
(Show $K = \{m \in \omega \mid u \in \omega$ whenever $u \subset m$ and $u = n \in \omega | n \in u\}$ is inductive.)

(xii) Any nonempty $S \subset \omega$ contains a *minimal element* $s$ (*i.e.* for each $t \in S$ $- \{s\}$, $S \in t$).

**Proof.** It suffices to show that $s = \bigcap S$ satisfies all our requirements: Clearly $s = \{n \in \omega | n \in s\}$. It is also easy to see that $s \in S$. (Suppose $s \notin S$. Then, for each $t \in S$, we get that $t \neq s$ because $s \subset t$ for every $t \in S$. Therefore, $s^* \subset t$ for each $t \in S$, which implies that $s^+ \subset \bigcap S = s$, a contradiction.) Obviously $s$ is the unique minimal element of $S$.

The arithmetic of the natural numbers can be entirely based on (i), (iii), (vi) and (vii). These imply the Peano Axioms of Arithmetic, which are

(a) zero is a natural number,
(b) every natural number has a successor,
(c) zero is not a successor of any natural number,
(d) natural numbers having the same successor are equal,
(e) a set, which contains zero and also the successor of every number in it, contains all natural numbers.

A *sketch of Arithmetic in* $\omega$. We define the operation of addition in $\omega$ by the following inductive procedure:

(1) $0 + 0 = 0$.
(2) Suppose we have already defined $j + k$ for $j, k \in n$ with $n \neq 0$. Then we let $t + n = (t + n^-)^+$, $n + t = (n^- + t)^+$ for any $t \in n$.

By induction, it is easy to verify that

(3) For every $j, k \in \omega$, $j + k \in \omega$.
(4) For every $j, k \in \omega$, $j + k = k + j$.
(5) For every $i, j, k \in \omega$, $(i + j) + k = i + (j + k)$.
(6) For every $j \in \omega$, $j + 0 = 0 + j = j$.

The multiplication operation in $\omega$ is treated in a similar fashion:

(7) $0 \times 0 = 0$
(8) Suppose we have already defined $j \times k$ for $j, k \in n$. Then, we let $n \times t = (n^- \times t) + t$, for any $t \in n$.

By induction, it is easy to verify that

(9) For every $j$, $k \in \omega$, $j \times k \in \omega$.
(10) For every $j$, $k \in \omega$, $j \times k = k \times j$.
(11) For every $i$, $j$, $k \in \omega$, $(j \times k) \times i = j \times (k \times i)$.
(12) For every $j \in \omega$, $j \times 1 = 1 \times j = j$.
(13) For every $i$, $j$, $k \in \omega$, $i \times (j + k) = i \times j + i \times k$.

As the reader well knows, it is customary to let

$$0^+ \equiv 1 (0^+ = 0 \cup \{0\} = \{0\}),$$
$$1^+ \equiv 2 (1^+ = \{0\} \cup \{\{0\}\} = \{0, \{0\}\}),$$
$$2^+ \equiv 3 (2^+ = \{0, \{0\}, \{0, \{0\}\}\}),$$
$$3^+ \equiv 4, 4^+ = 5, \ldots, n^+ = n + 1, \text{ and so on.}$$

The reader has certainly seen *definitions by induction* (namely, definitions of sequences $s : \omega \to X$, for some set $X$, with $s(n^+)$ depending on $s(n)$ in some prescribed manner (for example, $u_0 = 1$ and $u_{n+1} = u_2^+ + 2$, for every $n \in \omega$ (and also definitions of special restrictions of given functions $f : X \to X$ (for example, starting with a point $a \in X$, let $v_0 = a$, $v_1 = f(a), \ldots, v_{n+1} = f(u_n)$ for all $n \in \omega$). The principle of induction does guarantee, easily, that there can be at most one such sequence $\{u_n\}$ or $\{v_n\}$ satisfying the stated conditions, but it does not guarantee their *existence*. The existence of $\{u_n\}$ really depends on a better understanding of the "structure" of the set $\omega$. (We will sketch the existence of $\{u_n\}$ in Section 10.) Similarly, the existence of $\{v_n\}$ depends on our exact understanding of the notion of a function and the so-called Recursion Theorem, which we will state and prove later in Section 9.

*It is common practice* to let $\mathbf{N} = \omega - \{0\}$ and to call $\mathbf{N}$ the set of *natural numbers*. One can live with this!

## 0.8 Simple Cartesian Products

For any two elements $a$, $b$ of any set $C$ we let

$$(a, b) = \{\{a\}, \{a, b\}\}$$

(note that $(a, b)$ is a well-defined set, since $(a, b) \in \mathfrak{p}(\mathfrak{p}(C))$); $(a, b)$ is called the *ordered pair* of $a$ and $b$, with $a$ the *first element* and $b$ the *second element* of $(a, b)$— (this terminology makes sense, since

$$a \neq b \text{ implies } (a, b) = \{\{a\}, \{a, b\}\} \neq \{\{b\}, \{a, b\}\} = (b, a)).$$

It follows that, for any sets $A$ and $B$

$$A \times B = \{(a, b) | a \in A, b \in B\}$$

is a well-defined set, since, by the definition of ordered pair and the Axiom of Power Sets, $A \times B \subset \mathfrak{p}(\mathfrak{p}(A \cup B))$. The set $A \times B$ is called the *Cartesian Product* of the sets $A$ and $B$.

## 0.9   Relations

For any sets $X$ and $Y$, any $R \subset X \times Y$ is called a *relation between $X$ and $Y$* or from $X$ to $Y$. For convenience, we let $(x, y) \in R \equiv xRy$. We also let

$$dam\, R = \{x \in X | (x, y) \in R \text{ for some } y \in Y\},$$

$$rng\, R = \{y \in Y | (x, y) \in R \text{ for some } x \in X\}.$$

If $A \subset X$, we let

$$R|A = \{(a, y) \in R | a \in A\} = \{(a, y) | a \in A, aRy\}.$$

(Then $R|A$ is a relation between $A$ and $Y$ and is called the *restriction of $R$ to $A$*.) If $X = Y = dam\, R$, one simply says that *$R$ is a relation* on $X$. A relation $R$ on $X$ is said to be

- (i)  *reflexive*, if $xRx$ for each $x \in X$,
- (ii)  *symmetric*, if $xRy$ implies $yRx$,
- (iii)  *antisymmetric*, if $xRy$ and $yRx$ imply $x = y$,
- (iv)  *transitive*, if $xRy$ and $yRz$ imply $xRz$.

An *order* $\leq$ on a set $X$ is a reflexive, antisymmetric and transitive relation on $X$. A *total order* $\leq$ on a set $X$ is an order on $X$ such that, for every $x$, $y \in X$, either $x \leq y$ or $y \leq x$.

We say that $(X, \leq)$ is an *ordered set* (resp. *totally ordered set*) provided that $X$ is a set and $\leq$ is an order (resp. a total order) on $X$.

An *equivalence relation $R$* on a set $X$ is a relation which is reflexive, symmetric and transitive. (Equivalence relations are *extremely important!*)

## 0.10   Functions

For any sets $X$ and $Y$, a *function $f$ between $X$ and $Y$* is a relation between $X$ and $Y$ which satisfies

- (i)  $dom\, f = X$.
- (ii)  $xfy, xfz \Rightarrow y = z$.

It is customary to denote a function $f$ between $X$ and $Y$ by

$$f : X \to Y \text{ or } X \xrightarrow{f} Y$$

and to let

$$xfy \equiv y = f(x).$$

Given a function $f : X \to Y$, we call $X$ the *domain* of $f$ and $\{f(x) | x \in X\}$ the *range* of $f$. We also let

$$\text{graph of } f \equiv gr(f) = \{(x, f(x)) | x \in X\}.$$

For each $a \in Y$, we let $c_a : X \to Y$ denote the constant function defined by $c_a(x) = x$, for each $x \in X$. The identity function from $X$ to $X$ is denoted by $i_X$.

If $f : X \to Y$ and $A \subset X$, we denote by $f|A$ the function (!) $f|A : A \to Y$, defined by $(f|A)(a) = f(a)$, for each $a \in A$.

A function $f : X \to Y$ is said to be

   (i) *onto* or *surjective* if for each $y \in Y$ there exists $x \in X$ such that $y = f(x)$,
   (ii) *one-to-one*, 1–1 or *injective* if $y = f(x)$ and $y = f(w)$ imply $x = w$,
   (iii) *bijective* or *a one-to-one correspondence* if $f$ is 1–1 and onto.

We will let

$$f : X \to Y \text{ is onto} \equiv f : X \twoheadrightarrow Y.$$

Each function $f : X \to Y$ produces a function $f : \mathfrak{p}(X) \to \mathfrak{p}(Y)$ (one should use a different symbol for this new function, but it is fun not to) defined by

$$f(A) = \{f(x) | x \in A\},$$

for each $A \subset X$.

Given functions $X \xrightarrow{f} Y$, $Y \xrightarrow{g} Z$, we let $g \circ f : X \to Z$ be the function, defined by $g \circ f(x) = g(f(x))$ for each $x \in X$(!) and we call $g \circ f$ the *composite function* of $f$ and $g$. Note that

$$g \circ f = \{(x, z) | (x, y) \in f \text{ and } (y, z) \in g, \text{ for some } y \in Y\}.$$

To keep a promise, we now state and prove the very useful Recursion Theorem.

**Recursion Theorem.** If $X$ is a set, $f : X \to X$ a function and $a \in X$, then there exists a function $v : \omega \to X$ such that $v(0) = a$ and $v(n^+) = f(v(n))$ for all $n \in \omega$.

**Proof.** We limit ourselves to a sketch, leaving the easy details for the reader. Let

$$C = \{A \subset \omega \times X | (0, a) \in A \text{ and } (n^+, f(x)) \in A \text{ if } (n, x) \in A\}.$$

Clearly $C \neq \emptyset$, since $\omega \times X \in C$. Let $v = \bigcap C$. It follows that $v \in C$. Therefore, it remains to show that $v$ is a function: Let

$$K = \{n | (n, x) \in v, \text{ for at most one } x\}.$$

To show that $v$ is a function, simply show that $K$ is inductive. (For example, to show that $0 \in K$: Suppose $(0, a) \in v$ and $(0, c) \in v$ with $c \neq a$. Let $B = \omega \times X - \{(0, c)\}$. It is immediate that $B \in C$ and therefore that $v \subset B$, which implies that $(0, c) \notin v$, a contradiction.)

## 0.11    Sequences

A *sequence in the set A* is a function from the set $\mathbf{N}$ of positive integers (or from $\omega$) to the set $A$. It is customary to denote a sequence $f : \mathbf{N} \to A$ by enumerating its range: $f(1), f(2), \ldots$ To emphasize the attachment of each $n \in \mathbf{N}$ to some element $f(n)$ of $A$, it is customary to denote $f(n)$ by $a_n$ thus indicating that the integer $n$ is attached to the element $a_n$ of $A$. In this fashion, one simply says that $a_1, a_2, \ldots$ is a sequence in the set $A$. For convenience, we let

$$a_1, a_2, \ldots \equiv \{a_n\}_{n=1}^{\infty} \equiv \{a_n\} \equiv \{a_n\}_n.$$

A *subsequence* of the sequence $\{a_n\}$ is any sequence $\{b_k\}$ such that

(i) $\{b_k | k = 1, 2, \ldots\} \subset \{a_n | n = 1, 2, \ldots\}$,
(ii) There exists a function $t : \{k | k \in \mathbf{N}\} \to \{n | n \in \mathbf{N}\}$ such that $k_1 < k_2$ implies $t(k_1) < t(k_2)$ and for every $n$ there exists $k_n$ such that $t(k_n) > n$, (*i.e.*, $t$ is *increasing* and $\{t(k) | k \in \mathbf{N}\}$ is *cofinal* in $\mathbf{N}$).

(Note that, letting $t(k) = n_k$, we then get that

$$\{b_k\}_k \equiv \{a_{t(k)}\} \equiv \{a_{n_k}\}_k,$$

the last notation being very popular.)

A sequence $a_n$ is *finite* provided there exists $n \in \mathbf{N}$ such that $a_n = a_{n+1} = a_{n+2} \ldots$ Otherwise, $\{a_n\}$ is said to be *infinite*. (Note that if $\{a_n\}$ is a finite sequence then its range is finite. But a sequence may have finite range and not be a finite sequence; for example, the sequence $1, 2, 1, 2, 1 \ldots$ has range $\{1, 2\}$ but is not finite.)

To keep another promise, let us show that it is valid to define sequences with domain $\omega$ recursively. (In particular, *there exists a unique* function $u : \omega \to \omega$ such that $u_0 = 1$ and $u_{n+1} = u_n^2 + 2$, for each $n \in \omega$.)

**2. Proposition.** Let $X$ be a set which satisfies the following conditions:

(i) $u_0 \in X$,
(ii) $u_{n+1} = R(u_n) \in X$ for $n = 1, 2, \ldots$ (*i.e.*, each $u_{n+1}$ is chosen depending on $u_n$ by some explicit rule $R$).

Then there exists a unique function $f : \omega \to X$ such that $f(0) = u_0$ and $f(n) = u_n$ for $n = 1, 2, \ldots$.

**Proof.** Let us call a subset $A$ of $\omega \times X$ a *string* if it has the property

$$(^*)(0, u_0) \in A \text{ and } ((n + 1), R(x)) \in A \text{ whenever } (n, x) \in A.$$

Let $\mathcal{C}$ be the collection of all strings. Note that $\mathcal{C} \neq \emptyset$, since $\omega \times X \in \mathcal{C}$. Finally, let $f = \bigcap \mathcal{C}$. It is easy to see that $f$ is a function (simply show that $K = \{n | (n, x) \in f$ for at most one $x\}$ is inductive). It is also easy to see that $f \in \mathcal{C}$, from which it follows that $f$ is the unique required function.

## 0.12 Indexing Sets

Sometimes it is very convenient to attach a "name" to each element of a collection $\mathcal{C}$ of sets. This is easily done by picking a set and a function $f : \mathcal{I} \to \mathcal{C}$ (certainly this can always be done—for example, let $\mathcal{I} = \mathcal{C}$ and $f : \mathcal{I} \to \mathcal{C}$ be the identity function). Indeed, we *cannot require* that the function $f$ be *injective*, because we may not be free to choose the indexing set. Then, for each $i \in \mathcal{I}$, the set $f(i) \in \mathcal{C}$ has the *name* $f(i)$. Since it is clear that the function is not important but only the knowledge of which set $C \in \mathcal{C}$ corresponds to $i \in \mathcal{I}$, it is customary to let

$$f(i) = C_i,$$

thus indicating that the set $C_i$ came from the collection $\mathcal{C}$ and has been given the *name i*. We then say that

$$\mathcal{C} = \{C_i | i \in \mathcal{I}\} \equiv \{C_i\}_{i \in \mathcal{I}},$$

and $\mathcal{I}$ is called an *indexing set* for $\mathcal{C}$. With this new language, if $\mathcal{U} = \{U_\alpha\}_{\alpha \in \Lambda}$,

$$\bigcup \mathcal{U} \equiv \bigcup \{U_\alpha | \alpha \in \Lambda\} \equiv \bigcup_{\alpha \in \Lambda} U_\alpha,$$

$$\bigcap \mathcal{U} \equiv \bigcap \{U_\alpha | \alpha \in \Lambda\} \equiv \bigcap_{\alpha \in \Lambda} U_\alpha,$$

For indexing sets which'are subsets of the integers, the notations

$$\bigcup_{n \geq j} A_n = \bigcup_{n=j}^{\infty} A_n, \qquad \bigcup_{n \geq 1} A_n = \bigcup_{n} A_n, \ldots$$

are acceptable, their meaning being clearly understood (for example, $\bigcup_{j=n}^{\infty} A_j \equiv \bigcup \{A_n | n$ is an integer and $n \leq j < \infty\}$).

## 0.13 Important Formulas

Let $X$ be a set, $\{A_1, \ldots, A_n\}$ any finite family of sets and $\{A_\alpha\}_{\alpha \in A}$ any nonempty family of sets. Then

(i) $X - \bigcap_{\alpha \in \Lambda} A_\alpha = \bigcup_{\alpha \in \Lambda} (X - A_\alpha)$,

(ii) $X - \bigcup_{\alpha \in \Lambda} A_\alpha = \bigcap_{\alpha \in \Lambda} (X - A_\alpha)$,

(iii) $X - \bigcup_{i=1}^{n} A_i = X - A_1 - A_2 - \ldots - A_n$, it being understood that $X - A_1 - \ldots - A_{k+1} \equiv (X - A_1 - \ldots - A_k) - A_{k+1}$ (definition by induction!)

(iv) $X \cap (\bigcup_{\alpha \in \Lambda} A_\alpha) = \bigcup_{\alpha \in \Lambda} (X \cap A_\alpha)$,

(v) $X \cup (\bigcap_{\alpha \in \Lambda} A_\alpha) = \bigcap_{\alpha \in \Lambda} (X \cup A_\alpha)$,

(vi) $X - (A_1 - A_2) \supset (X - A_1) - A_2$, $X - (A_1 - A_2) \neq (X - A_1) - A_2$, generally. (Contrast this with (iii). Sometimes, parentheses do make a difference!)

## 0.14   Inverse Functions

Given a function $f : X \to Y$, we define a function $f^{-1} : \mathfrak{p}(Y) \to \mathfrak{p}(X)$ by

$$f^{-1}(B) = \{x \in X | f(x) \in B\},$$

for each $B \subset Y$. Note that if $B \subset Y$ and $B \cap \{f(x) | x \in X\} = \emptyset$, then $f^{-1}(B) = \emptyset$. Also note that $f^{-1}|\{\{y\} : y \in Y\}$ can be thought of as a function from $Y$ to $X$ if and only if $f$ is bijective, in this case, we define $f^{-1} : Y \to X$ by letting

$$f^{-1}(y) = (f^{-1}|\{\{y\} : y \in Y\})(\{y\}),$$

for each $y \in Y$; of course, we should use another symbol for the new function $f^{-1} : Y \to X$, but that would not make things any clearer.

## 0.15   More Important Formulas

For any function $f : X \to Y$, the following are valid:

   (i) $ff^{-1}(B) \subset B$, for each $B \subset Y$ (we should write $f(f^{-1}(B))$ instead of $ff^{-1}(B)$, but we tend to get confused with too many parentheses while others get upset by too few; these and other discordances certainly give life to the politics of mathematics),

  (ii) If $f$ is onto then $ff^{-1}(B) = B$, for each $B \subset Y$,

 (iii) $A \subset f^{-1}f(A)$, for each $A \subset X$,

 (iv) For any family $\{B_\alpha\}_{\alpha \in \Lambda}$ of subsets of $Y$,

$$f^{-1}(\textstyle\bigcup_{\alpha \in \Lambda} B_\alpha) = \bigcup_{\alpha \in \Lambda} f^{-1}(B_\alpha),$$
$$f^{-1}(\textstyle\bigcap_{\alpha \in \Lambda} B_\alpha) = \bigcap_{\alpha \in \Lambda} f^{-1}(B_\alpha),$$

  (v) $f^{-1}(Y - B) = X - f^{-1}(B)$,

$$f^{-1}(B) = X - f^{-1}(Y - B), \text{ for each } B \subset Y,$$

 (vi) For any family $\{A_\alpha\}_{\alpha \in \theta}$ of subsets of $X$,

$$f(\textstyle\bigcup_{\alpha \in \theta} A_\alpha) = \bigcup_{\alpha \in \theta} f(A_\alpha),$$
$$f(\textstyle\bigcap_{\alpha \in \theta} A_\alpha) \subset \bigcap_{\alpha \in \theta} f(A_\alpha),$$
$$f(\textstyle\bigcap_{\alpha \in \theta} A_\alpha) \neq \bigcap_{\alpha \in \theta} f(A_\alpha), \text{ generally.}$$

## 0.16   Partitions

A *partition* of a set $X$ is a collection $\mathfrak{p}$ of subsets of $X$ such that $\bigcup \mathfrak{p} = X$ and $\mathfrak{p}$ is pairwise disjoint (*i.e.*, for each $A$, $B \in \mathfrak{p}$, with $A \neq B$, $A \cap B = \emptyset$). If $A \subset X$, it is convenient to denote the partition $\{A\} \cup \{\{x\} | x \in X - A\}$ of $X$ by $X/A$.

    Whenever convenient, we make no distinction between the singleton $\{x\}$ and the element $x \in X$. For example, we let

$$X/A = \{A\} \cup \{x | x \in X - A\}.$$

## 0.17 Equivalence Relations, Partitions and Functions

These concepts are, undoubtedly, the essence of Mathematics. It is therefore crucial that the reader have no second thoughts about the following (except for the bad English): "Equivalence Relations" generate "Partitions" generate "Functions" generate "Equivalence Relations."

Here is how it happens:

(i) Let $R$ be an equivalence relation on $X$. For each $x \in X$, let $[x] = \{y \in X | yRx\}$. ($[x]$ is called the *R-equivalence class of x*.) Then, $\{[x] | x \in X\}$ is a partition of $X$. For convenience, let $X/R = \{[x] : x \in X\}$.

(ii) Let $\mathfrak{p}$ be a partition of $X$. Define a relation $f$ from $X$ to $\mathfrak{p}$ by

$$x f s \text{ iff } x \in s.$$

Then $f$ is a function (*i.e.*, $f : X \to \mathfrak{p}$ and $f(x) = the\ P \in \mathfrak{p}$ *which contains* $x$).

(iii) Let $f : X \to Y$ be a function. Define a relation $R$ on $X$ by

$$x_1 R x_2 \text{ if and only if } f(x_1) = f(x_2).$$

Then $R$ is an equivalence relation on $X$, such that $[x] = f^{-1}f(x)$, for each $x \in X$.

## 0.18 General Cartesian Products

Let $a = \{A_\alpha\}_{\alpha \in \Lambda}$ be a family of sets. The *Cartesian product* of the family $\{A_\alpha\}_{\alpha \in \Lambda}$ is the set

$$\left\{ f : \Lambda \to \bigcup_{\alpha \in \Lambda} A_\alpha | f(\beta) \in A_\beta \text{ for each } \beta \in \Lambda \right\}$$

and is denoted by $\prod_{\alpha \in \Lambda} A_\alpha$. (Note that, to be precise, we should emphasize that $\prod_{\alpha \in \Lambda} A_\alpha \subset \mathfrak{p}(\Lambda \times \bigcup_{\alpha \in \Lambda} A_\alpha)$.) If $a$ is finite, it is easy to prove, from the aforementioned axioms of set theory, that $\prod_{\alpha \in \Lambda} A_\alpha$ is nonempty. *If it is not finite, it has been proved that, from the aforementioned axioms of set theory, one can neither deduce that $\prod_{\alpha \in \Lambda} A_\alpha$ is empty nor that it is nonempty.* (*We need more axioms!*)

Given the sets $A_1, \ldots, A_n$, with $n$ a positive integer, and letting $S = \{l, 2, \ldots, n\}$, it is customary to let

$$\prod_{i=1}^{n} A_i \equiv \prod_{i \in S} A_i.$$

Also, extending the notion of an ordered pair to the notion of an "ordered $n$-tuple" $(a_1, \ldots, a_n)$, in some convenient way, it is customary to let

$$\prod_{j=1}^{n} A_i = \{(a_1, \ldots, a_n) | a_1 \in A_1, \ldots, a_n \in A_n\}$$

it being clearly understood that $(a_1, \ldots, a_n)$ corresponds, in a one-to-one fashion, to the function $f : S \to \bigcup_{i=1}^{n} A_i$ such that $f(i) = a_i \in A_i$, for $i = 1, \ldots, n$. The reader is well-advised to think in terms of functions rather than $n$-tuples, since functions impose no limitations on the *size* of the index set $\Lambda$; on the other hand, to think of tuples which have more elements than the integers may cause headaches.

In some instances, it is very convenient to replace $f : \Lambda \to \bigcup_{\alpha \in \Lambda} A_\alpha$ by its image $f(\Lambda)$ and to give it a more familiar appearance; namely, we let

$$f \equiv f(\Lambda) \equiv (f(\alpha))_\alpha \equiv (a_\alpha)_\alpha,$$

it being understood that $(a_\alpha)_\alpha$ represents the function $f : \Lambda \to \bigcup_{\alpha \in \Lambda} A_\alpha$ such that $f(\alpha) = a_\alpha$, for each $\alpha \in \Lambda$. By no means, under any circumstances, try to attach some order to $(a_\alpha)_\alpha$ *since none is implied*. Of course, in case $\Lambda = \omega$, we can experience the sensation of order, by letting

$$(a_n)_n \equiv (a_0, a_1, \ldots).$$

For each family $\{A_\alpha\}_{\alpha \in \Lambda}$ of sets and $\beta \in \Lambda$, we define the *$\beta$-projection* $\prod_\beta :$ $\prod_{\alpha \in \Lambda} A_\alpha \to A_\beta$ by letting $\prod_\beta(f) = f(\beta) \in A_\beta$ for each $f \in \prod_{\alpha \in \Lambda} A_\alpha$. (Clearly, each $\prod_\beta$ is a function (!).)

## 0.19   The Sixth Axiom (Axiom of Choice)

Here is that fascinating axiom you've waited for so patiently.

VI. (*Axiom of Choice*). Let $\{A_\alpha\}_{\alpha \in \Lambda}$ be a non-empty family of non-empty sets. Then $\prod_{\alpha \in \Lambda} A_\alpha$ is non-empty. (Essentially, this says that, given a non-empty collection of non-empty sets, one may form a new set by picking one element from each set.)

It seems quite safe to say that this axiom has generated more research in the Foundations of Mathematics than any other axiom of any mathematical discipline. Roughly, it can be said that its significance was tested in three different ways.

(i) *Equivalent Axioms:* Today there are various equivalent forms of the axiom of choice. Some of these are far from being obviously equivalent to the axiom of choice and some have attained great significance in mathematics—especially, the *Well-Ordering Theorem* and *Zorn's Lemma* (neither is a theorem, of course; further ahead, we will state these *axioms* without proving their equivalence to the Axiom of Choice).

(ii) *Consistency.* It is relatively easy to prove that, from the first five axioms of set theory one cannot obtain a contradiction or a false statement by logical reasoning. In the late 1930s, Gödel proved that from the six axioms of set theory one cannot obtain a contradiction or a false statement by logical reasoning (*i.e.*, the Gödel-Bernays axiom system is *consistent*).

(iii) *Independence.* Answering the question of consistency led to another question: Is the axiom of choice a consequence of the other five axioms of Gödel-Bernays by logical reasoning? (*i.e.*, is the axiom of choice a theorem?) Recently—1964—P. J. Cohen proved that the answer is *no*. Essentially, he did this by constructing a collection $\mathcal{K}$ of sets which satisfies the first five axioms of Gödel-Bernays and also the negation of the axiom of choice (*i.e.*, these statements are true in $\mathcal{K}$). $\mathcal{K}$ is called a *model* for these axioms.

The reader should not overlook the obvious: While the independence results of Cohen are, by any means, tremendously impressive, they do not constitute the last word on these matters; after all, *is the model acceptable as the best and only description of the world around us?* (Is it even a good imitation of it?—Playing with models does pose some challenges!)

## 0.20   Well-Orders and Zorn

We will now present two axioms which are equivalent to the Axiom of Choice. Traditionally, the first is known as the Well-Ordering Theorem and the second is known as Zorn's Lemma. Clearly, neither is a theorem—this only indicates the confusion that has surrounded these matters.

Before stating these axioms we need a few preliminaries: A total order $\leq$ on a set $E$ is said to be a *well-order on $E$* provided that, for each $A \subset E$, there exists $m \in A$ such that $m \leq b$, for each $b \in A$. The element $m$ is said to be a *minimal element of $A$* (on $E$ with respect to $\leq$).

For every set $\mathcal{K}$ and any order $\leq$ on $\mathcal{K}$ ($\leq$ can be total or not) we say that

(i) $\mathcal{N} \in \mathcal{K}$ is a $\leq$-*nest* if $\leq |\mathcal{N} \times \mathcal{N}$ is a total order in $\mathcal{N}$ (*i.e.*, for every $x_1, x_2 \in \mathcal{N}$, either $x_1 \leq x_2$ or $x_2 \leq x_1$),

(ii) $q \in \mathcal{K}$ is said to be a $\leq$-*maximal* element of $\mathcal{K}$ if $\{x \in \mathcal{K} | q \neq x, q \leq x\} = \emptyset$ (*i.e.*, there exists no $x \in \mathcal{K} - \{q\}$ such that $q \leq x$). Analogously, we define $\leq$-*minimal* elements.

(iii) Given $S \subset \mathcal{K}$, we say that $q \in \mathcal{K}$ is a $\leq$-*upper bound* of $S$ provided that $s \leq q$, for each $s \in S$. Analogously, we define $\leq$-*lower bound*.

*Well-Ordering Axiom.* Given any set $A$, there exists a well-order in $A$.

*Zorn's (Lemma) Axiom.* If $\leq$ is an order (total or not) on a set $A$ such that every $\leq$-nest in $A$ has an $\leq$-upper bound in $A$, then $A$ has at least one $\leq$-maximal element.

Please pay attention to what the Well-Ordering Axiom really says: Undoubtedly, it *does not say* that any order on a set $A$ is a *well-order*. It simply says that *there exists* a well-order on any given set; it does not even say if there exists only one or if we can *construct* one. For example, note that, if $A = \{-n | n \in \mathbf{N}\}$ with the usual order (*i.e.*, $-m \leq -n$ if and only if $n \leq m$), then $\leq$ is not a well-order on the set $A$, even though the usual order on $\omega$ is a well-order. Finally, even though the Well-Ordering Axiom guarantees the existence of a Well!-Order on the set $\mathfrak{p}(\omega)$, no one has ever constructed one on $\mathfrak{p}(\omega)$, and no one ever will with these axioms.

    **Transfinite Induction Theorem.** Let $W$ be a well-ordered set whose first element is 0. Suppose $P(\alpha)$ is a proposition indexed by $\alpha \in W$. If $P(0)$ is true and the truth of $P(\alpha)$, for $\alpha < \beta$, implies the truth of $P(\beta)$, then $P(\gamma)$ is true, for all $\gamma \in W$.

**Proof.** Routine, by contradiction.

## 0.21   Yet More Important Formulas

Let $\{A_\alpha\}_{\alpha \in \Lambda}$, $\{B_\alpha\}_{\alpha \in \Lambda}$ and $\{C_\beta\}_{\beta \in \theta}$ be non-empty families of sets. Then

    (i) $\bigcup\{A_\alpha \times C_\beta | \alpha \in \Lambda, \beta \in \theta\} = (\bigcup_{\alpha \in \Lambda} A_\alpha) \times (\bigcup_{\beta \in \theta} C_\beta)$,

    (ii) $\bigcup\{A_\alpha \times B_\alpha | \alpha \in \Lambda\} \neq (\bigcup_{\alpha \in \Lambda} A_\alpha) \times (\bigcup_{\alpha \in \Lambda} B_\alpha$, (For example, let $A_1 = \{1\}$, $A_2 = \{2\}$, $B_1 = \{3\}$, $B_2 = \{4\}$. Then $A_1 \times B_1 \cup A_2 \times B_2 = \{(1,3),(2,4)\}$, while $(A_1 \cup A_2) \times (B_1 \cup B_2) = \{(1,3),(1,4),(2,3),(2,4)\}$. Do contrast (ii) with (i).)

    (iii) $\bigcap\{A_\alpha \times B_\alpha | \alpha \in \Lambda\} = (\bigcap_{\alpha \in \Lambda} A_\alpha) \times (\bigcap_{\alpha \in \Lambda} B_\alpha)$.

    (iv) $\bigcap\{A_\alpha \times C_\beta | \alpha \in \Lambda, \beta \in \theta\} = (\bigcap_{\alpha \in \Lambda} A_\alpha) \times (\bigcap_{\beta \in \theta} C_\beta)$.

## 0.22   Cardinality

The axiom of infinity poses some interesting questions: Given two infinite sets $A$ and $B$, can we compare their *sizes*? Does $A$ have *more* elements than $B$? Anyway, what does one mean by the *size* of an infinite set? This brings us back to finite sets and a careful analysis of what we really mean by *counting the elements of a finite set*—unquestionably, what we do when we count the apples in a basket-full of apples is to establish a *one-to-one* correspondence between the apples and other objects (generally, the natural numbers starting with 1). It is thus clear that the concept of one-to-one correspondence is really the key to success, when it comes to *sizes* of sets:

    *Cardinals.* Given two sets $A$ and $B$, we let

$$A \preceq B$$

denote that there exists an injective function from $A$ to $B$. We also let

$$A \prec B \equiv A \preceq B, \text{ but not } B \preceq A.$$

    Two sets $A$ and $B$ are said to be *equipollent* provided that there exists a one-to-one correspondence between $A$ and $B$. (It is commonly said that $A$ and $B$ have the *same cardinal number*; we avoid this language because we do not really have enough ammunition to convince the reader that it makes sense to talk about the "*same*" cardinal number—this certainly presupposes that one already knows what a "cardinal number" is, and an honest treatment of this concept would lead us too far astray.) For convenience, we let

$$A \text{ is equipollent with } B \equiv A \# B.$$

It is obvious that

1. For any set $A$, $A \# A$,
2. If $A = B$, then $B \# A$,
3. $A \# B$, $B \# C$ implies that $A \# C$ (the composition of two bijections is a bijection!).

Therefore, $\#$ is an equivalence relation on any given family of sets. Therefore, if we could really talk about the "*collection* $\mathcal{U}$ *of all sets*" one could then define the cardinal number of any set $A$ as the family of all sets which are equipollent with A (*i.e.*, the equivalence class of $A$ in $\mathcal{U}$). How appealing! And how deceiving!

**3. Proposition.** There is no set $\mathcal{U}$ such that any set is an element of $\mathcal{U}$ (*i.e.*, there is no *universe*).

**Proof.** Suppose there exists a set $\mathcal{U}$ of all sets. Then, by the Axiom of Extension, $\mathcal{U}$ is unique. Either $\mathcal{U} \in \mathcal{U}$ or $\mathcal{U} \notin \mathcal{U}$. The assertion that $\mathcal{U} \notin \mathcal{U}$ leads to an immediate contradiction. Therefore, we must have that $\mathcal{U} \in \mathcal{U}$. Now, let $B = \{x | x \in \mathcal{U}, x \in x\}$ and note that $B \neq \emptyset$, since $\mathcal{U} \notin \mathcal{U}$. We have two cases to consider.

Case 1. $B = \mathcal{U}$: Then the set $\emptyset \notin \mathcal{U}$, since $\emptyset \notin \emptyset$.

Case 2. $B \neq \mathcal{U}$: Let $A = \mathcal{U} - B$. Then $A$ is a nonempty *set* and $A = \{x | x \in U, x \notin x\} = \{x | x \notin x\}$. But we already know, from Section 3, that the assumption that $\{x | x \notin x\}$ is a *set* leads to a contradiction. This completes the proof.

The following result states the obvious. The proof we give here, which is neither ours nor the original proof of Schröder or Bernstein, is remarkably simple.

**Schröder-Bernstein Theorem.** $X \preceq Y, Y \preceq X \Rightarrow X \# Y$.

**Proof.** Let $f : X \to Y$ and $g : Y \to X$ be injections. Observe that we could immediately finish the proof if we knew that there exists some $A \subset X$ such that

$$B = f(A), g(Y - B) = X - A,$$

for then we would simply define $h : X \to Y$ by

$$h(x) = \begin{cases} f(x), & x \in A, \\ g^{-1}(x), & x \in X - A, \end{cases}$$

it being obvious that $h$ is a bijection from $X$ to $Y$.

Therefore, to complete the proof, we will show that actually there exists $A \subset X$ such that

$$B = f(A), \ g(Y - B) = X - A.$$

First, let $M = X - g(Y - f(X))$ and let us show that $f(Y - f(M)) \supset X - M$: Simply note that, because $f$ and $g$ are 1-1,

$$g(Y - f(M)) = g(Y - f[X - g(Y - f(X))])$$
$$= g(Y) - gf[X - g(Y - f(X))]$$
$$\supset g(Y) - gf(X) = f(Y - f(X))$$
$$= X - [X - g(Y - f(X))] = X - M.$$

Now, let $\mathfrak{a} = \{S \subset X | g(Y - f(S)) \supset X - S\}$ and note that $\mathfrak{a} \neq \emptyset$, since $M \in \mathfrak{a}$. So, let $A = \bigcup \mathfrak{a}$ and $B = f(A)$. It remains for us to show that $g(Y - B) = X - A$. This will be done in two parts.

$g(Y - B) \supset X - A$: Note that $g(Y - B) \supset g(Y - f(S)) \supset X - S$ for each $S \in \mathfrak{a}$. Therefore, $g(Y - B) \supset \bigcup\{X - S | S \in \mathfrak{a}\} = X - \bigcap \mathfrak{a} = X - A$.

$g(Y - B) \subset X - A$: Suppose not. Then there exists $z \in g(Y - B)$ such that $z \notin X - A$. Letting $A_* = A - \{z\}$, we get that

(i) $A_* \subset A$ and $A_* \neq A$, since $z \in A$,
(ii) $g(Y - f(A_*)) \supset X - A_*$, since $X - A_* = (X - A) \cup \{z\}$, $z \in g(Y - B) \subset g(Y - f(A_*))$ and $g(Y - B) \supset X - A$.

Since (i) and (ii) contradict the definition of $A$, our supposition is false. Hence $g(Y - B) \subset X - A$, which completes the proof.

The next result is a masterpiece and a shocker.

**Cantor's Theorem.** If $X$ is a non-empty set then $X \prec \mathfrak{p}(X)$.

**Proof.** Clearly $f : X \to \{\{x\} | x \in X\}$, defined by $f(x) = \{x\}$ for every $x \in X$, is a bijective function between $X$ and a subset of $\mathfrak{p}(X)$.

Suppose there exists a bijective map $h : X \to \mathfrak{p}(X)$. Let $A = \{x \in X | x \notin h(x)\}$. Since $A \in \mathfrak{p}(X)$ and $h$ is onto, there exists $a \in X$ such that $h(a) = A$. Either $a \in A$ or $a \notin A$. But $a \in A$ implies that $a \notin h(a) = A$, a contradiction. And $a \notin A$ implies that $a \in h(a) = A$, another contradiction. So, our assumption that there exists a bijective map $h : X \to \mathfrak{p}(X)$ has led us to an impossibility; therefore, $X \prec \mathfrak{p}(X)$.

Let us say that a set $S$ is *countable* provided that $S$ is equipollent to some subset of $\omega$ (possibly $\omega$ itself). A set $T$ is said to be *uncountable* provided that $T$ is not countable. (*By Cantor's Theorem, there exist uncountable sets.*) Naturally, we say that a set $F$ is finite provided that $F$ is equipollent with some natural number $n$. A set $K$ is infinite provided that $K$ is not finite.

The following elementary, but quite useful, results are stated without proof.

(i) If the natural numbers $m$ and $n$ are equipollent then $m = n$. (Note that it suffices to show that $K = \{n | n$ is not equipollent to any $m \in n\}$ is inductive.)

(ii) If $T \subset \omega$ is infinite then $T \# \omega$.

(iii) If $A \subset B$ then either $A \prec B$ or $A \# B$.

(iv) There exists a surjective function $f : \omega \twoheadrightarrow X$ iff $X$ is nonempty and countable.

(The "if" part is really obvious. For the "only if" part, simply pick a function $g \in \prod_{x \in X} f^{-1}(x)$, by the Axiom of Choice. It follows that $g$ is an injection from $X$ to $\omega$, which implies that $X$ is countable. We could disguise our use of the Axiom of Choice by simply defining a function $g : X \to \omega$, with $g(x) = $ some $n \in f^{-1}(x)$, for each $x \in X$ note that $f^{-1}(x) \neq \emptyset$, because $f$ is onto—but this is really just a disguise.)

**4. Proposition.** There exists a sequence $\{A_n\}$ of infinite subsets of $\omega$ such that

(a) $A_n \cap A_m = \emptyset$ whenever $m \neq n$,

(b) $\omega = \bigcup_{n=0}^{\infty} A_n$.

**Proof.** (First of all, let us emphasize that there are many ways of proving this result. The one we use here fits the preceding development.) Let

$$A_0 = \{0\} \cup \{2j + 1 | j = 0, 1, \ldots\}.$$

and

$$A_n = \{2^n(2j + 1) | j = 0, 1, \ldots\}, \text{ for } n = 1, 2, \ldots.$$

It is easy to see that $\omega = \bigcup_{n=0}^{\infty} A_n$ (clearly we only need to show that each even number is in some $A_n$; but it is easy to show that each even number $k$ has the form $k = 2^j m$ with $m$ odd; since, for $m \in A_0 - \{0\}$, $2^j m \in A_j$, this does the trick).

To show that $A_n \cap A_m = \emptyset$ whenever $n \neq m$, note that, assuming $m < n$,

$$2^n(2j + 1) = 2^m(2k + 1) \text{ implies } 2^{n+1}j + 2^n = 2^{m+1}k + 2^m$$

$$\text{implies } 2^{m+1}(2^{n-m}j - k) = 2^{n-m} - 1$$

$$\text{implies } 2(2^{n-m}j - k) = 2^{n-m} - 1,$$

which implies that an even number equals an odd number, a contradiction. Therefore,

$$2^n(2j + 1) = 2^m(2k + 1) \text{ implies } m = n, \text{ which implies that}$$

$$A_n \cap A_m = \emptyset \text{ whenever } n \neq m.$$

This result immediately gives us an extremely useful result, which we will call the "Cbl-Cbl Theorem", "Cbl" standing for "Countable".

**CBL-CBL Theorem.** The union of a *countable* collection of *countable* sets is countable.

**Proof.** Let $\mathcal{C}$ be a countable collection of countable sets. Using 22(iv), let $\ell : \omega \twoheadrightarrow \mathcal{C}$ be a surjection. Say $\ell(n) = C_n$, for each $n \in \omega$. Making use of the sets $A_n$ of Proposition 4, the hypothesis that each $C \in \mathcal{C}$ is countable and of 22(ii), let

$$\gamma_n : A_n \twoheadrightarrow C_n$$

be a surjection. Finally, define $f : \omega \to \bigcup \mathcal{C}$, by letting

$$f(i) = \gamma_n(i) \text{ for each } n \in \omega.$$

It follows that $f$ is surjective and, therefore, that $\bigcup \mathcal{C}$ is countable.

**5. Proposition.** For any finite family $A_1, \ldots, A_n$ of countable sets, $\prod_{i=1}^{n} A_i$ is countable.

**Proof.** Because of induction, it suffices to prove that, if $A$ and $B$ are countable then $A \times B$ is countable: Simply observe that

$$A \times B = \bigcup_{b \in B} (A \times \{b\}),$$

and therefore that $A \times B$ is a countable union of countable sets.

The Cbl-Cbl Theorem and Proposition 4 strongly support the conjecture that *any Cartesian product of countably many countable sets is countable.* After all, from the viewpoint of equipollence, what difference can there be between countable products and countable unions? Surprise! Surprise!

**6. Theorem.** For any sequence $\{A_n\}_{n=1}^{\infty}$ of infinite sets, $\prod_{i=1}^{\infty} A_i$; is uncountable.

**Proof.** Without loss of generality, we assume that each $A_i = \omega$. Since, by Cantor's Theorem, we already know that $\mathfrak{p}(\omega)$ is uncountable, it suffices to show that there exists an onto function $f : \prod_{i=1}^{\infty} \omega \to \mathfrak{p}(\omega)$. (Simply let $f((x_1, x_2, \ldots)) = \{f(x_i) | i = 1, 2, \ldots\}$, for each $(x_1, x_2, \ldots) \in \prod_{i=1}^{\infty} \omega$, and use Axiom of Choice to show that $f$ is onto.)

# Chapter 1

# Metric and Topological Spaces

The definition of continuity of a real-valued function of a real variable is certainly well known to the reader, and yet there are subtleties about it that may have passed your scrutiny.

Let us therefore look at its many equivalent forms as a prelude for the reasons behind the definition of metric and topological spaces.

So, let $E^1$ be the real line and, for $a, b \in E^1$, let $]a, b[ = \{x \in E^1 | a < x < b\}$ and $[a, b] = \{x \in E^1 | a \leq x \leq b\}$; also let $f : E^1 \mapsto E^1$ be a function. By definition, $f$ is continuous at $p$ provided that

for each $\varepsilon > 0$ there exist $\delta > 0$ such that

$$|x - p| < \delta \text{ implies } |f(x) - f(p)| < \varepsilon,$$

equivalently,

for each $\varepsilon > 0$ there exist $\delta > 0$ such that

$$p - \delta < x < p + \delta \text{ implies } f(p) - \varepsilon < f(x) < f(p) + \varepsilon,$$

equivalently,

for each $\varepsilon > 0$ there exist $\delta > 0$ such that

$$x \in ]p - \delta, p + \delta[ \text{ implies } f(x) \in ]f(p) - \varepsilon, f(p) + \varepsilon[,$$

equivalently,

for each $\varepsilon > 0$ there exist $\delta > 0$ such that

$$f\big(]p - \delta, p + \delta[\big) \subset ]f(p) - \varepsilon, f(p) + \varepsilon[,$$

equivalently,

for each $]a, \beta[$ containing $f(p)$ there exists

$$]b, \sigma[ \text{ containing } p \text{ such that } f\big(]b, \sigma[\big) \subset ]a, \beta[.$$

No doubt, only the last equivalence may require some thoughts. However, the observation that

> for each $]a, \beta[$ containing $f(p)$ there exists
>
> $\varepsilon > 0$ such that $]f(p) - \varepsilon, f(p) + \varepsilon[ \subset ]a, \beta[$.

and

> for each $]b, \sigma[$ containing $p$ there exists $\delta > 0$ such that $]p - \delta, p + \delta[ \subset ]b, \sigma[$

should make it all clear.

We have therefore gone from the definition of continuity which requires the notion of absolute value to an equivalent definition which requires that the image under $f$ of "small" sets containing $p$ be "small" sets containing $f(p)$.

The reader must also be aware that in the proofs of the key theorems on continuous real-valued functions (namely, if $f : E^1 \to E^1$ and $g : E^1 \to E^1$ are functions which are continuous at $p$, then $f + g$, $f \circ g$ and $\dfrac{f}{g}$ (assuming $g(p) \neq 0$) are continuous at $p$) the only properties of the absolute value which were used were the ones which we now summarize (we use the traditional, even though imprecise, notation).

*The absolute value in $E^1$ is a function, whose domain is $E^1 \times E^1$ and whose range is contained in $E^1$, which satisfy the following:*

(i) $|x - y| \geq 0$
(ii) $|x - y| = 0$ iff $x = y$,
(iii) $|x - y| = |y - x|$,
(iv) $|x - y| \leq |x - z| + |z - y|$.

The reader can also easily check that the proofs of these theorems, in terms of the last equivalent forms of the definition of continuity of the functions $f$, $g$ at $p$, require only the following fact about open intervals of the real line:

*Given any point $q \in E^1$ and any two open intervals $N_1$, $N_2$, with $q \in N_1 \cap N_2$, there exists some open interval $N_3$ such that*

$$q \in N_3 \subset N_1 \cap N_2.$$

It is now clear that there is too much about the real line that has nothing to do with the continuity of functions. Therefore, the need to eliminate all that superfluous structure from the context of continuity is clear.

## 1.1   Metrics and Topologies

**1.   Definition.**   A *metric space* $(M, \rho)$ is a set $M$ together with a function $\rho : M \times M \to E^1$ such that for all $x, y, z \in M$,

(i) $\rho(x, y) \geq 0$,
(ii) $\rho(x, y) = 0$ iff $x = y$,

(iii) $\rho(x,y) = \rho(y,x)$ (symmetry),

(iv) $\rho(x,z) \le \rho(x,y) + \rho(y,z)$ (triangle inequality).

**2. Definition.** A *topological space* $(X,\tau)$ is a set $X$ together with a family $\tau$ (called, *topology*) of subset of $X$ such that

(i) $a \subset \tau$ implies $\bigcup a \in \tau$,

(ii) $\mathcal{F} \subset \tau$ and $\mathcal{F}$ is finite implies $\bigcap \mathcal{F} \in \tau$,

(iii) $\emptyset \in \tau$ and $X \in \tau$.

The reader should immediately observe that a topology $\tau$ *is closed with respect to unions* (*i.e.*, a union of element of $\tau$ is an element of $\tau$) *and also closed with respect to finite intersections*. The apparent absurdity of requiring that $\tau$ be closed *with respect to* unions but *only with respect to finite* intersections can easily be explained by the observation that *infinite* intersections of open intervals may not be open intervals (for example, $\bigcap_{n=1}^{\infty}] - \frac{1}{n}, \frac{1}{n}[= \{0\})$ and by the observation that to allow degenerate intervals $[p-0, p+0] = \{p\}$ in the definition of continuity of a function $f : E^1 \to E^1$ would yield that all functions $h : E^1 \to E^1$ are continuous, thus rendering the all-too-important concept of continuity of real-valued functions completely useless.

From Definition 2(i) it follows immediately that the collection $\mathcal{B}$ of open intervals of $E^1$ is not a topology for $E^1$ (cf. $]0,1[ \cup ]3,4[$ is not an open interval). But the collection

$$\mu = \left\{ \bigcup a \, | \, a \subset \mathcal{B} \right\}$$

(*i.e.*, the collection of unions of open intervals; note that $\emptyset \in \mu$, because $\emptyset \in \mathcal{B}$ and $\bigcup \emptyset = \emptyset$) is indeed a topology (!) for $E^1$. Henceforth, we will call the topology $\mu$ *the Euclidean topology*. This technique of producing the topology $\mu$ from the much simpler subfamily $\mathcal{B}$ (note that $\mathcal{B} \subset \mu$) is far too important to be dismissed.

**3. Definition.** Let $(X,\tau)$ be a topological space. A subfamily $\mathcal{B}$ of $\tau$ is called a *base* for topology $\tau$ iff each $U \in \tau$ is a union of elements of $\mathcal{B}$ (*i.e.*, $U = \bigcup \mathcal{U}$, for some $\mathcal{U} \subset \mathcal{B}$).

**4. Theorem.** A collection $\mathcal{B}$ of sets is a base for some topology $\tau$ on $X = \bigcup \mathcal{B}$ iff for each pair $N_1, N_2$ of elements of $\mathcal{B}$, and for each $p \in N_1 \cap N_2$, there exists some $N_3 \in \mathcal{B}$ such that $p \in N_3 \subset N_1 \cap N_2$; furthermore, $\tau = \{\bigcup a \, | \, a \subset \mathcal{B}\}$.

**Proof.** The *if* part: Clearly $\tau = \{\bigcup a \, | \, a \subset \mathcal{B}\}$ is closed with respect to unions; $\emptyset, X \in \tau$, and $\mathcal{B} \subset \tau$ ($B \in \mathcal{B}$ implies $\{B\} \subset \mathcal{B}$, and therefore that $B = \bigcup\{B\} \in \tau$). Therefore, we only need to show that $\tau$ is closed with respect to finite intersections, for which it clearly suffices to show that if $\bigcup a_1, \bigcup a_2 \in \tau$ then $(\bigcup a_1) \cap (\bigcup a_2) \in \tau$. Without loss of generality, since $\emptyset \in \tau$, we assume that $(\bigcup a_1) \cap (\bigcup a_2) \ne \emptyset$. Then, for each $x \in (\bigcup a_1) \cap (\bigcup a_2)$ there exist $A_1 \in a_1 \subset \mathcal{B}$

and $A_2 \in \mathcal{U}_2 \subset \mathcal{B}$ such that $x \in A_1 \cap A_2$; therefore, there exists $A_x \in \mathcal{B}$ such that $x \in A_x \subset A_1 \cap A_2 \subset (\bigcup \mathcal{U}_1) \cap (\bigcup \mathcal{U}_2)$; this shows that $\bigcup \mathcal{U}_1) \cap (\bigcup \mathcal{U}_2)$ is a union of elements of $\mathcal{B}$ (the $A_x$'s) and therefore $(\bigcup \mathcal{U}_1) \cap (\bigcup \mathcal{U}_2) \in \tau$.

The *only if* part: Suppose $\mathcal{B} \subset \tau$ is a base for the topology $\tau$. Clearly $\tau = \{\bigcup \mathcal{U} | \mathcal{U} \subset \mathcal{B}\}$. Furthermore, if $p \in N_1 \cap N_2$ with $N_1, N_2 \in \mathcal{B} \subset \tau$, then $p \in N_1 \cap N_2 \in \tau$; consequently, since $N_1 \cap N_2 \in \tau$ is a union of elements of $\mathcal{B}$, there exists $N_3 \in \mathcal{B}$ such that $p \in N_3 \subset N_1 \cap N_2$.

**5. Corollary.** If $\mathcal{S}$ is an arbitrary family of sets, then the family $\mathcal{B}$ of all finite intersections of elements of $\mathcal{S}$ is a base for a topology $\tau$ on $\bigcup \mathcal{S}$; furthermore, $\tau = \{\bigcup \mathcal{U} | \mathcal{U} \subset \mathcal{B}\} \supset \mathcal{B} \supset \mathcal{S}$.

**Proof.** We only need to show that if $x \in B_1 \cap B_2$ with $B_1, B_2 \in \mathcal{B}$, then there exists $B_3 \in \mathcal{B}$ such that $x \in B_3 \subset B_1 \cap B_2$: Say $x \in (\bigcap \mathcal{F}_1) \cap (\bigcap \mathcal{F}_2)$ with $\mathcal{F}_1, \mathcal{F}_2$ being finite subcollections of $\mathcal{S}$. Then $\mathcal{F}_3 = \mathcal{F}_1 \cap \mathcal{F}_2$ is a finite subcollection of $\mathcal{S}$ such that $x \in \bigcap \mathcal{F}_3 \subset (\bigcap \mathcal{F}_1) \cap (\bigcap \mathcal{F}_2)$.

Corollary 5 provides a very simple way of manufacturing a topology for the union of any collection $\mathcal{S}$ of sets. This technique is underlined by the following definition.

**6. Definition.** Let $(X, \tau)$ be a topological space. a subfamily $\mathcal{S}$ of $\tau$ is a *subbase* for topology $\tau$ iff the family $\mathcal{B} = \{\bigcap \mathcal{F} | \mathcal{F} \subset \mathcal{S} \text{ and } \mathcal{F} \text{ is finite}\}$ is a base for $\tau$. The topology $\tau$ is said to *be generated* by $\mathcal{S}$.

The reader should check that

$$\mathcal{S} = \{]a, +\infty[ \, | \, a \in E^1\} \cup \{] - \infty, b[ \, | b \in E^1\}$$

is a subbase for the Euclidean topology of $E^1$.

## 1.2    Time out for Notation

To facilitate the shop-talk concerning the study of metric and topological spaces we need a shorthand language. So, here it is.

    A. Let $(M, d)$ be a metric space.

        (i) For each $x \in M$ and $\varepsilon > 0$, $B(x, \varepsilon) = \{y \in M | d(x, y) < \varepsilon\}$ ($S(x, \varepsilon) = \{y \in M | d(x, y) = \varepsilon\}$) is called an $(x, \varepsilon)$-ball (an $(x, \varepsilon)$-sphere)).

       (ii) The metric $d$ on $M$ is called *bounded* if there exists $s > 0$ such that $d(x, y) \leq s$ for all $x, y \in M$.

      (iii) The set $A \subset M$ is said to be *bounded with respect to d* if there exists $s > 0$ such that $d(x, y) \leq s$ for all $x, y \in A$.

      (iv) For each $A \subset M$ and $B \subset M$,

$$\rho(A, B) = \inf\{d(a, b) | a \in A, b \in B\}$$

is called *the distance between* $A$ and $B$. For $p \in M$, we let $\rho(\{p\}, A) = \rho(p, A)$.

(v) For each $A \subset M$, $\sup\{d(x,y)|x,y \in A\}$ is called *the diameter of* $A$ and is *denoted by* diam $A$.

B. Let $(X, \tau)$ be a topological space.

  (i) The elements of $\tau$ are called $\tau$-*open sets* or just *open* sets, when no confusion is possible.

  (ii) A set $N \subset X$ is a $\tau$-*neighborhood*, or just *neighborhood* or nbhd, of $p \in X$ (or $P \subset X$) provided that there exists $U \in \tau$ such that $p \in U \subset N$ (or $P \subset U \subset N$).

  (iii) The complement of an open set is called a *closed* set ($A$ closed iff $(X - A) \in \tau$).

  (iv) For every $A \subset X$, the set $A^{-} = \bigcap\{B \subset X | A \subset B$ and $B$ is closed$\}$ is called the *closure* of $A$. (Whenever convenient, $A^{-} \equiv \bar{A}$).

  (v) For every $A \subset X$, the set $A^{0} = \bigcup\{U \subset A | U \in \tau\}$ is called the *interior* of $A$.

  (vi) For every $A \subset X$, the set $\partial A = A^{-} - A^{0}$ is called the *boundary* of $A$.

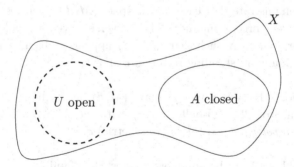

C. *Euclidean Spaces:* For each $n \in \mathbf{N}$, we let $E^{n}$ denote $n$-Euclidean space (*i.e.*, $E^{1}$ is the space of real numbers and, for $n \geq 2$, $E^{n} = \{(x_1, \ldots, x_n)|x_1, \ldots, x_n \in E^{1}\}$). We also let

$$B^{n} = \left\{(x_1, \ldots, x_n) \in E^{n}|x_1^2 + \cdots + x_n^2 \leq 1\right\}, \text{ for } n \in \mathbf{N}$$

and

$$S^{n} = \left\{(x_0, \ldots, x_n) \in E^{n+1}|x_0^2 + \cdots + x_n^2 = 1\right\}, \text{ for } n \in \omega.$$

$B^{n}$ is called the (*Euclidean*) $n$-*ball* and $S^{n}$ is called the (*Euclidean*) $n$-*sphere*. Clearly, $S^{n} = \partial B^{n+1}$. It is customary to let $I = \{x \in E^{1} | 0 \leq x \leq 1\}$. For convenience, we let $\bar{x} = (x_1, \ldots, x_n)$ and $|\bar{x}| = \sqrt{\sum_{i=1}^{n} x_i^2}$, or $|\bar{x}| = \sqrt{\sum_{i} x_i^2}$ whenever the number of coordinates is clear from the context.

The following very elementary facts are given without proof.

**7. Proposition.** Let $(X, \tau)$ be a topological space. Then

   (i) Finite unions and arbitrary intersections of families of closed subsets of $X$ are again closed subsets of $X$.

   (ii) For each $A \subset X$, $A^-$ is closed, $A^- \supset A$ and $(A^-)^- = A^-$. If $A$ is closed then $A^- = A$.

   (iii) For each $A \subset X$, $A^0$ is open $(A^0 \in \tau)$, $A^0 \subset A$ and $(A^0)^0 = A^0$. If $A$ is open then $A^0 = A$.

   (iv) $A \subset B$ implies $A^- \subset B^-$ and $A^0 \subset B^0$.

   (v) For each $A \subset X$, $\partial A$ is closed, $\partial A = A^- \cap (X - A)^- = \partial(X - A)$ and $A^- = A^0 \cup \partial A$.

   (vi) For each $A \subset X$, $A^0 = X - (X - A)^-$ and $A^- = X - (X - A)^0$.

   (vii) The set $A \subset X$ is closed iff $A \supset \partial A$; $A$ is open iff $A \cap \partial A = \emptyset$.

   (viii) For each $A \subset X$, $\partial(\partial A) = \partial A$.

   (ix) For each $A, B \subset X$, $(A \cup B)^- = A^- \cup B^-$ (see ex. 21).

## 1.3  Metrics Generate Topologies

Below, we demonstrate that each metric space $(M, d)$ generates a topological space $(M, \tau_d)$ in a very precise manner. The converse (given a topological space $(X, \tau)$, is there a metric $d$ on $X$ such that $\tau_d = \tau$?) appears to be a never-ending problem, even though many outstanding results are known.

**8. Definition.** For every metric space $(M, d)$, let $\tau_d$ be the topology generated the collection $S$ of all $(x, \varepsilon)$-balls $B(x, \varepsilon)$, with $x \in M$ and $\varepsilon > 0$. The topology $\tau_d$ is called the *topology generated by* (the metric) $d$.

**9. Lemma.** Let $(M, d)$ be any metric space. The family

$$S = \{B(x, \varepsilon) | x \in M, \varepsilon > 0\}$$

is actually a base for the topology $\tau_d$.

**Proof.** Because of Theorem 4, we only to show that if $p \in B(x, \varepsilon_1) \cap B(y, \varepsilon_2)$ then there exist $B(z, \varepsilon_3)$ such that $p \in B(z, \varepsilon_3) \subset B(x, \varepsilon_1) \cap B(y, \varepsilon_2)$: Since $d(x, p) < \varepsilon_1$ and $d(y, p) < \varepsilon_2$, then $d(x, p) = \varepsilon_1 - \delta_1$ and $d(y, p) = \varepsilon_2 - \delta_2$ with $\delta_1 > 0$ and $\delta_2 > 0$. Let $\delta = \min(\delta_1, \delta_2)$. By the triangle inequality, we get that

$$p \in B(p, \delta) \subset B(x, \varepsilon_1) \cap B(y, \varepsilon_2).$$

   The fact that each metric space generates an unique (!) topological space is the main reason why the study of topological spaces is far more intensive than the study of metric spaces. After all, whatever is valid for topological spaces must be valid for the topological spaces generated by metric spaces.

**10. Definition.** A topological space $(X, \tau)$ is *metrizable* provided that there exists a metric $d$ on $X$ such that $\tau = \tau_d$.

In general, topological spaces are not metrizable, even when these appear to be very simple (see ex. 27).

**11. Proposition.** Let $(M, d)$ be a metric space. Then

(i) $(B(x, \varepsilon))^- \subset \{y \in M | d(x, y) \le \varepsilon\}$. (Equality may not hold.)
(ii) $\bigcup \{B(y, \delta) | y \in B(x, \varepsilon)\} \subset B(x, \varepsilon + \delta)$. (Equality may not hold.)
(iii) $\partial[B(x, \varepsilon)] \subset S(x, \varepsilon)$. (Equality may not hold.)
(iv) $A \subset M$ is closed iff $d(p, A) = 0$ implies that $p \in A$.
(v) $p \in A^-$ iff $d(p, A) = 0$.
(vi) $A \subset M$ is bounded iff there exists $s > 0$ such that $A \subset B(a, s)$, for each $a \in A$.

## 1.4 Continuous Functions

Since all this is a result of our preoccupation with continuous real-valued functions, it is about time to see how far we have come.

**12. Definition.** Let $(X, \tau)$ and $(Y, \delta)$ be topological spaces and $f : X \to Y$ (or $f : (X, \tau) \to (Y, \delta)$) be a function. The function $f$ is said to be *continuous* at $p \in X$ (with respect to $\tau$ and $\delta$, of course) provided that for each $\sigma$-neighborhood $V$ of $f(p)$ there exists a $\tau$-neighborhood $U$ of $p$ such that $f(U) \subset V$. The function $f$ is said to be *continuous* on $X$ provided that it is continuous at each point of $X$.

The following result should convince and assure the reader that we have not modified the "usual" concept of contiuity of a real-valued function in the minutest detail.

**13. Lemma.** Let $(X, d)$ and $(Y, \rho)$ be metric spaces. A function $f : (X, \tau_d) \to (Y, \tau_\rho)$ is continuous at $p \in X$ if and only if

(a) For each $\varepsilon > 0$, there exists $\delta > 0$ such that $d(x, p) < \delta$ implies $\rho(f(x), f(p)) < \varepsilon$.

**Proof.** This follows immediately from definition of the subbase for the topology generated by a metric and from Lemma 9.

**14. Continuity Theorem.** Let $(X, \tau)$ and $(Y, \sigma)$ be topological spaces and $f : X \to Y$ be a function. Then the following statements are equivalent (see ex. 13):

(i) $f$ is continuous,

(ii) $f^{-1}(U) \in \tau$ for each $U \in \sigma$,

(iii) $f^{-1}(B)$ is $\tau$-closed for each $\sigma$-closed set $B$,

(iv) $f(A^-) \subset f(A)^-$ for each $A \subset X$,

(v) $f^{-1}(B)^- \subset f^{-1}(B^-)$ for each $B \subset Y$.

(vi) If $\mathcal{S}$ is a subbase for $\sigma$, then $f^{-1}(S) \in \tau$ for each $S \in \mathcal{S}$.

**Proof.** The scheme of the proof will be (i) implies (ii) implies (iii) implies (iv) implies (v) implies (i) and (ii) iff (vi)

(i) implies (ii): Let $U \in \sigma$. Then, for each $x \in F^{-1}(U)$ there exists some $N_x \in \tau$ such that $x \in N_x$ and $f(N_x) \subset U$, which implies that $N_x \subset f^{-1}(U)$. Therefore $f^{-1}(U) = \bigcup\{N_x | x \in f^{-1}(U)\} \in \tau$. (It is important to that $f^{-1}(U) = f^{-1}(U \cap f(X))!$).

(ii) implies (iii): Let $B$ be $\sigma$-closed. Then $X - f^{-1}(B) = f^{-1}(Y - B)$. Since $(Y - B) \in \sigma$ we then get that $f^{-1}(B)$ is $\tau$-closed.

(iii) implies (iv): Let $A \subset X$. Then, by (iii), $f^{-1}[f(A)^-]$ is $\tau$-closed and contains $A$. Therefore $A^- \subset f^{-1}[f(A)^-]$ which implies that $f(A^-) \subset ff^{-1}[f(A)^-] \subset f(A)^-$.

(iv) implies (v): Let $B \subset Y$ and let $A = f^{-1}(B)$. Then, by (iv), $f[f^{-1}(B)^-] \subset [ff^{-1}(B)]^- \subset B^-$ which implies that $f^{-1}(B)^- \subset f^{-1}(f[f^{-1}(B)^-]) \subset f^{-1}(B^-)$.

(v) implies (i): Let $p \in X$ and let $f(p) \in U \in \sigma$. Then, by (v), $f^{-1}(Y-U)^- \subset f^{-1}(Y - U)$ which implies that $f^{-1}(Y - U)^- = f^{-1}(Y - U)$. Therefore, $p \in [X - f^{-1}(Y - U)] \in \tau$ and $f(X - f^{-1}(Y - U)) = U$.

(ii) implies (vi): Obvious.

(i) implies (ii): Straightforward, since the topology $\sigma$ consist of unions of finite intersections of elements of $\tau$ and $f^{-1}$ commutes with unions and intersections.

**15. Corollary.** If $X \xrightarrow{f} Y \xrightarrow{g} Z$ and $f$ and $g$ are continuous, then $g \circ f$ is continuous.

**Proof.** Immediate from Theorem 14(ii), because

$$(g \circ f)^{-1}(U) = f^{-1}(g^{-1}(U)), \text{ for every } U \subset Z.$$

**16. Corollary.** Let $f(X, \tau) \to (Y, \mu)$ be one-to-one and onto. If $f$ and $f^{-1}$ are continuous then $\tau = \{f^{-1}(V) | V \in \mu\}$ and $\mu = \{f(U) | U \in \tau\}$

**Proof.** Note that $f$ continuous implies $\tau \supset \{f^{-1}(V) | V \in \mu\}$, and $f^{-1}$ continuous implies $\mu \supset \{f(U) | U \in \tau\}$. But $\mu \supset \{f(U) | U \in \tau\}$ implies $\tau \subset \{f^{-1}(V) | V \in \mu\}$, and $\tau \supset \{f^{-1}(V) | V \in \mu\}$ implies $\mu \subset \{f(U) | U \in \tau\}$.

What Corollary 16 really says, is that, given its hypothesis, the only difference between $(X, \tau)$ and $(Y, \mu)$ is one of *color*—each $x \in X$ is colored $f(x)$ and put into $Y$ and each $y \in Y$ is colored $f^{-1}(y)$ and put into $X$.

**17. Definition.** Two spaces $(X, \tau)$ and $(Y, \mu)$ are said to be homeomorphic (symbolically, $X \simeq Y$) if there exists a one-to-one and onto function $f : X \to Y$ which is *bicontinuous* (*i.e.*, $f$ and $f^{-1}$ are continuous). The function $f$ is called a *homeomorphism*.

From the proof of Corollary 16, the reader can easily conclude that homeomorphisms have the properties of sending *open sets to open sets* and *closed sets to closed sets*. Functions with these properties play an important role in various topological constructions.

**18. Definition.** An onto function $f : (X, \tau) \to (Y, \mu)$ is

(a) *open* iff $f(U) \in \mu$, for each $U \in \tau$
(b) *closed* iff $f(A)$ is closed for each $A \subset X$.

## 1.5 Subspaces

The proof that (i) implies (ii) in Theorem 14 shows that, in discussing the continuity of a function $f : (X, \tau) \to (Y, \sigma)$, one only cares about the image $f(X)$ of $X$ ($Y - f(X) = \emptyset$ iff $f$ is onto) and the collection

$$\sigma_* = \{U \cap f(X) | U \in \sigma\}$$

which is easily seen to be a topology (!) on $f(X)$. Indeed, it is always the case that, for any topological space $(X, \tau)$ and subset $A$ of $X$,

$$\{U \cap A | U \in \tau\}$$

is a topology for $A$ (obvious!). This leads to the following definition.

**19. Definition.** Let $(X, \tau)$ be any topological space and $A \subset X$. Then the topology $\{U \cap A | U \in \tau\}$ is denoted by $\tau | A$ and is called *the relative topology* on $A$. The space $(A, \tau | A)$ is called a (topological) *subspace* of the space $(X, \tau)$.

Of course, metric spaces also have (metric) subspaces, in an obvious fashion: For any metric space $(X, d)$ and $A \subset X$ we assign to $A$ the metric $d_A = d | A \times A$ (remember $d$ is a function!). Then $(A, d_A)$ is called a *metric subspaces* of $(X, d)$.

## 1.6 Comparative Topologies

Any set $X$ with more than one element has at least two distinct topologies: The *indiscrete* topology $\mathcal{T} = \{\emptyset, X\}$ and the *discrete* topology $\mathcal{D} = \{A | A \subset X\}$. Corollary 5 implies that $X$ has many other topologies.

Given two topologies $\tau, \sigma$ on a set $X$, $\tau$ is said to be *finer* than $\sigma$ than (and $\sigma$ is said to be coarser than $\tau$) whenever $\sigma \subset \tau$. It follows that $\mathcal{D}$ is the finest topology for $X$ and $\mathcal{T}$ is the coarsest topology for $X$.

The following result, even though trivial, is extremely useful.

**20. Lemma.** Two topologies $\tau_1, \tau_2$ on a set $X$ are equal iff there exist bases $\mathcal{B}_1$, $\mathcal{B}_2$ of $\tau_1, \tau_2$, respectively, such that

    (a) each element of $\tau_1$ is an union of elements of $\mathcal{B}_2$ (this shows that $\tau_1 \subset \tau_2$), and

    (b) each element of $\tau_2$ is an union of elements of $\mathcal{B}_1$ (this shows that $\tau_2 \subset \tau_1$).

## Chapter 1.  Exercises

1. Let $\mathcal{B}$ be a collection of sets. Show that, for each family $\{\mathcal{U}_\alpha\}_{\alpha \in \Lambda}$ of subcollections of $\mathcal{B}$,

$$\bigcup \left\{ \bigcup \mathcal{U}_\alpha \mid \alpha \in \Lambda \right\} = \bigcup \{U \mid U \in \mathcal{U}_\alpha\}, \text{ for some } \alpha \in \Lambda.$$

2. Let $\mathcal{S} = \{[a, +\infty[ \mid a \in E^1\} \cup \{] - \infty, b] \mid b \in E^1\}$. Then $\mathcal{S}$ is a subbase for the so-called *discrete topology* $\tau(D)$ on $E^1$. Describe a base $\mathcal{B}$ for $\tau(D)$.

3. Let $\mathcal{S} = \{[a, +\infty[ \mid a \in E^1\} \cup \{] - \infty, b[ \mid b \in E^1\}$. Then the collection $\mathcal{S}$ is a subbase for the so-called *"half-open interval topology"* $\tau_S$ on $E^1$. Show that $\mathcal{B} = \{[a, b[ \mid a, b \in E^1 \text{ and } a < b\}$ is a base for $\tau_S$. $((E^1, \tau_S)$ is called the *Sorgenfrey line*.)

4. Let $f : E^1 \to E^1$ be a function and $p$ a point of $E^1$. Prove that the following statements are equivalent:

    (a) For every $\varepsilon > 0$ there exists $\delta > 0$ such that $|x - p| < \delta$ implies $|f(x) - f(p)| < \varepsilon$.

    (b) For every $]\alpha, \beta[$ containing $f(p)$ there exists $]\mu, \sigma[$ containing $p$ such that $f(]\mu, \sigma[) \subset ]\alpha, \beta[$.

5. Let $A$ and $B$ be subspaces of $(X, \tau)$ with $A \subset B$. Then $A$ is a *subspace* of $B$.

6. Show that the intersection of any collection of topologies for a set $X$ is a topology for $X$. Is the union of two topologies for $X$ a topology for $X$? (see ex. 8).

7. Show that

    (a) $\mathcal{B} = \{[a, b[ \times [c, d[ | a, b, c, d \in E^1\}$ is a base for a topology $\gamma_s$ on $E^2$, which is different from the Euclidean topology for $E^2$. (Note that $[a, b[ \times [c, d[$ is a rectangle with the left and lower edges included, and the upper and right edges excluded.)

    (b) $A = \{(x, y) \in E^2 | y = -x\}$, with the topology $\gamma_s | A$, is a discrete space. $((E^2, \gamma_s)$ is called the *Sorgenfrey plane*.)

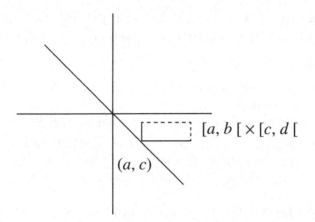

8. Let $X$ be a partially ordered set. Let $U_\ell(x) = \{y|y \leq x\}$ and $U_r(x) = \{y|x \leq y\}$. Show that

   (a) $\{U_\ell(x)|x \in X\}$ and $\{U_r(x)|x \in X\}$ are bases for topologies $\mathscr{I}_\ell, \mathscr{I}_r$ on $X$, respectively.

   (b) The discrete topology is the only topology on $X$ larger than $\mathscr{I}_\ell$ and $\mathscr{I}_r$.

9. Let $G$ be a subset of a topological space $(X, \tau)$. Show that $G \in \tau$ iff $G \cap A^- \subset \overline{G \cap A}$, for each $A \subset X$.

10. For any linearly ordered set $X$, let $\tau_0(X)$ be the topology with subbasis consisting of all subsets of $X$ of the form $\{x|x > a\}$ or $\{x|x < a\}$. Show that

    (a) $\tau_0(E^1)$ is the Euclidean topology.

    (b) If $A = \{0\} \cup \{x : |x| > 1\}$, then $\tau_0(E^1)|A \neq \tau_0(A)$.

11. Let $X$ be the set of all $(n \times n)$-matrices of real numbers. For each $(a_{ij}) \in X$ and $\varepsilon > 0$, let $B((a_{ij}), \varepsilon) = \{(b_{ij}) \in X| |a_{ij} - b_{ij}| < \varepsilon$, for all $i, j\}$. Show that

    (a) $\mathcal{B} = \{B((a_{ij}), \varepsilon)|(a_{ij}) \in X, \varepsilon > 0\}$ is a base for a topology $\tau$ on $X$.

    (b) $(X, \tau)$ is homeomorphic to the Euclidean space $E^{n^2}$.

12. A function $f : (X, \tau) \twoheadrightarrow (Y, \sigma)$ is closed iff, for each $B \subset Y$ and open $U$ containing $f^{-1}(B)$, there exists open $V \supset B$ such that $f^{-1}(V) \subset U$.

13. Let $f : (X, \tau) \to (Y, \sigma)$ be continuous functions and $A \subset X$. Show that $f|A : (A, \tau|A) \to (Y, \sigma)$ and $f|A : (A, \tau|A) \to ((f(A), \sigma)|f(A))$ are continuous functions.

14. Let $(E^1, \tau_h)$ be the Sorgenfrey line (see ex. 3) and define $f : (E^1, \tau_h) \to (E^1, \tau_h)$ by $f(x) = -x$. Is $f$ continuous?

15. Let $f : E^2 \to E^2$ be the function defined by $f(x) = x$ if $|x| \geq 1$ and $f(x) = 0$ if $|x| < 1$. Show that $f$ is a closed function which is not continuous.

16. Suppose $(X, \tau)$ is the union of two closed (open) subspace $A$ and $B$, $f_1 : (A, \tau|A) \to (Y, \sigma)$ and $f_2 : (B, \tau|B) \to (Y, \sigma)$ are continuous and $f_1(c) = f_2(c)$ for each $c \in A \cap B$. Define $f : (X, \tau) \to (Y, \sigma)$ by $f(x) = f_1(x)$ if $x \in A$ and $f(x) = f_2(x)$ if $x \in B$. Show that $f$ is continuous.

17. Show that $f : (X, \tau) \to (Y, \sigma)$ is open iff $f^{-1}(\partial B) \subset \partial f^{-1}(B)$, for each $B \subset Y$.

18. Let $\{A_\alpha | \alpha \in \Lambda\}$ be any family of subsets of a space $(X, \tau)$. Show that, if $\bigcup_\alpha A_\alpha^-$ is closed then $\bigcup_\alpha A_\alpha^- = \left(\bigcup_\alpha A_\alpha\right)^-$.

19. Let $A$ be a subset of a space $(X, \tau)$. Show that $\partial A = \emptyset$ iff $A$ is an open and closed subset of $X$.

20. Let $(M, \rho)$ be a metric space $p, q \in M$ and $A \subset M$. Show that

    (a) $\rho(q, A) \leq \rho(p, A) + \rho(q, p), \rho(p, A) \leq \rho(q, A) + \rho(p, q)$ (note that $\rho(q, a) \leq \rho(a, p) + \rho(q, p)$, for each $a \in A \dots$ ).

    (b) $|\rho(p, A) - \rho(q, A)| \leq \rho(p, q)$, since either $\rho(p, A) \leq \rho(q, A)$ or $\rho(q, A) \leq \rho(p, A)$.

    (c) Let $h : M \to E^1$ be defined by $h(x) = \rho(x, A)$. Show that $h$ is continuous. (Note that, for every $x, y \in M$ and $\varepsilon > 0$, $\rho(x, y) < \varepsilon$ implies that $|\rho(x, A) - \rho(y, A)| \leq \rho(x, y) < \varepsilon$.)

21. Let $(X, \tau)$ be a topological space and $A_1, \dots, A_n$ be subsets of $X$, and $n$ a positive integer. Show that

    (a) $A_1^- \cup \cdots \cup A_1^n (A_1 \cup \cdots \cup A_1)^-$. (Hint: Start with two sets $A$ and $B$ (see Prop. 7(ix)) and then use induction.)

    (b) $A_1^0 \cap \cdots \cap A_n^0 = (A_1 \cap \cdots \cap A_n)^0$. (Hint: For two subsets $A$ and $B$ of $X$, $A^0 \cap B^0) \supset (A \cap B)^0$, by Prop. 7(iv). Also, if $x \in A^0 \cap B^0$ then there exists an open neighborhood $U$ of $x$ such that $U \subset A$ and $U \subset B$ and thus $U \subset A \cap B$; that is, $x \in A^0 \cap B^0$ implies $x \in (A \cap B)^0$. Now apply induction!)

22. In $(E^1, \mu)$ (*i.e.*, the real line with the Euclidean topology), let $A$ be the set of rational numbers in $I$ and $B$ be the set of irrational numbers in $I$. Show that

    (a) $A^- \cap B^- = I$ and $(A \cap B)^- = \emptyset$.

    (b) $(A \cup B)^0 = ]0, 1[$ and $A^0 \cup B^0 = \emptyset$.

Let $(X, \tau)$ be a topological space. We say that

    (a) $D \subset X$ is *dense* in $X$ iff $D^- = X$.

    (b) $(X, \tau)$ is *second countable* iff there exists a countable base for $\tau$. $(X, \tau)$ is *first countable* if each $p \in X$ has a countable nbhd base.

    (c) $X, \tau$ is *separable* iff there exists a countable subset $D$ of $X$ such that $D$ is dense in $X$.

    (d) $(X, \tau)$ satisfies the *countable chain condition* (or has CCC) iff every pair-wise disjoint family $\mathcal{U}$ (*i.e.*, for every $U, V \in \mathcal{U}, U \cap V = \emptyset$) of open subsets of $X$ is countable.

    (e) $(X, \tau)$ is *Lindelöf* iff every open cover $\mathcal{U}$ of $X$ (*i.e.*, $\mathcal{U} \subset \tau$ and $\bigcup \mathcal{U} = X$) contains a countable subcover $\mathcal{C}$ (*i.e.*, $\mathcal{C} \subset \mathcal{U}$, $\mathcal{C}$ is countable $\bigcup \mathcal{C} = X$).

Two metrics $\rho, d$ for a set $X$ are said to be *equivalent* iff the topologies $\tau_\rho = \tau_d$.

23. Let $X$ be any uncountable set and $\tau = \{\emptyset\} \cup \{U \subset X | X - U \text{ is countable}\}$. Show that

(a) $(X, \tau)$ is a topological space.

(b) $(X, \tau)$ has CCC.

(c) $(X, \tau)$ is not separable.

24. Let $(X, d)$ be a metric space. Prove that

    (a) $(X, \tau_d)$ is separable implies $(X, \tau_d)$ is second countable. (Hint: Let $D = \{x_1, x_2, \ldots\}$ such that $D^- = X$. Show that $\{B(x_i, \frac{1}{n}|i, n = 1, 2, \ldots)\}$ is a base for $\tau_d$).

    (b) $(X, \tau_d)$ is second countable implies that $(X, \tau_d)$ is Lindelöf. This is valid for any topological space!

    (c) $(X, \tau_d)$ is Lindelöf implies $(X, \tau_d)$ is separable. (Hint: For each $n$, let $D_n$ be the set of centers of countably many balls with radius $1/n$, which covers $X$. Let $D = \bigcup_n D_n$ and show $D^- = X$.)

    (d) $(X, \tau_d)$ is separable implies $(X, \tau_d)$ has CCC.

    (e) $(X, \tau_d)$ has CCC implies $(X, \tau_d)$ is separable. (Hint: For each $n \in N$ let $D_n$ be the set centers of a maximal pair-wise disjoint collection of balls with radius $1/n$. Let $D = \bigcup_n D_n$ and show $D^- = X$.)

25. If $A$ is a dense subset of $(X, \tau)$ and $U \in \tau$ show that $U \subset (A \cap U)^-$.

26. Show that any Euclidean space $E^n$, with the topology generated by $S = \{\prod_{i=1}^n ]a_i, b_i[ \,|\, a_i, b_i \in E^1,$ for $i = 1, 2, \ldots, n\}$ is separable. (Hint: Use induction in a *rational* manner. $S$ is actually a base for this topology.)

27. Let $X$ be any uncountable set and choose a point $p \notin X$. Let $\hat{X} = X \cup \{p\}$. Show that

    (a) $\mathcal{B} = \{\{x\}|x \in X\} \cup \{\hat{X} - F|F$ is finite$\}$ is a base for a topology $\tau$ on $\hat{X}$.

    (b) $(\hat{X}, \tau)$ is not separable,

    (c) $(\hat{X}, \tau)$ is not second countable,

    (d) $(\hat{X}, \tau)$ does not have CCC,

    (e) $(\hat{X}, \tau)$ is Lindelöf (therefore, not metrizable, by ex. 24).

28. Let $(E^1, \tau_s)$ be the Sorgenfrey line (see ex. 3). Show that

    (a) $(E^1, \tau_s)$ is separable (consider the set of rational numbers).

    (b) $(E^1, \tau_s)$ is not second countable. (Any base for $\tau_s$ must contain some $[a, a + \delta[$, for each $a \in E^1$.)

    (c) $(E^1, \tau_s)$ is not metrizable.

29.    (a) Define $d : E^1 \times E^1 \to E^1$ by $d(x, y) = \dfrac{|x - y|}{1 + |x - y|}$.

        Show that $d$ is a metric. (Hint: The easiest way to show that $d$ satisfies the triangle inequality is by writing down what is wanted and simplifying it—it becomes obvious.)

    (b) Let $\Gamma_1 : E^1 \times E^1 \to E^1$ be defined by $\Gamma_1(x, y) = |x - y|$. Show that $\Gamma_1$ is a metric which is equivalent to $d$. (Hint: Recall Lemma 16.)

30. Show that, for metric spaces $(X, d)$, $(Y, \rho)$ and function $f : X \rightarrow Y$, the following are equivalent:

    (a) $f$ is continuous at the point $p \in X$,

    (b) For each sequence $\{x_n\}$ in $X$, with $\lim_n d(x_n, p) = 0$

        $\lim_n \rho(f(x_n), f(p)) = 0$.

# Chapter 2

# From Old to New Spaces

The reader is already familiar with various techniques of obtaining new structures from old familiar ones: For example, the complex numbers are obtained by forming all pairs $(a, b)$ of réal numbers, and even the addition and multiplication of complex numbers is based on the same operations for real numbers. Also, factor groups are obtained from two given groups $G, H$ and a homomorphism $h$ from $G$ onto $H$ (one then gets the factor group $G/Kerh$ which is isomorphic to $H$).

This *snail's approach* to scientific discovery is timeless and common to all research endeavors: Man, with his very limited wisdom has managed only to move from one structure to another a bit more complicated.

The reader should study what follows with this viewpoint in mind.

## 2.1 Product Spaces

Given a finite family $(X_i, \tau_i)$, $i = 1, 2, \ldots, n$, of topological spaces, our starting preoccupation is to give the set $\Pi_{i=1}^n X_i$ a "nice" topology $\tau$ *with respect to* which all projections (cf. 0.18) $\Pi_1, \ldots, \Pi_n$, are continuous. Certainly, if $\tau$ is the discrete topology, then all projections are continuous, but this is a useless topology, since *with respect to* it all functions with domain $\Pi_{i=1}^n X_i$ are continuous. Therefore, our real preoccupation must be to *find the coarsest topology for $\Pi_{i=1}^n X_i$ with respect to which all projections $\Pi_1, \ldots, \Pi_n$ are continuous*. This topology certainly must contain the collection

$$\Pi \mathcal{S} = \{\Pi_i^{-1}(U_i) | U_i \in \tau, i = 1, 2, \ldots, n\},$$

because of Theorem 1.14(ii).

A useful way of visualizing an element of $\Pi \mathcal{S}$ is to string all the spaces $X_i$ side-by-side

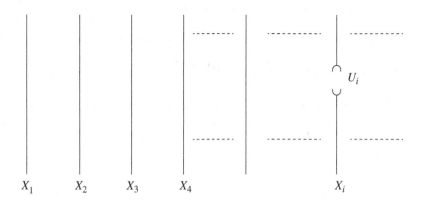

and to observe that $\pi_i^{-1}(U_i)$ consists of all functions in the product (tuples, if you wish) which assign to each $k \neq i$ any point of $X_k$ but must assign a point of $U_i$ to $i$. Then a basis element $\pi_{i_1}^{-1}(U_{i_1}) \cap \cdots \cap \pi_{i_j}^{-1}(U_{i_j})$

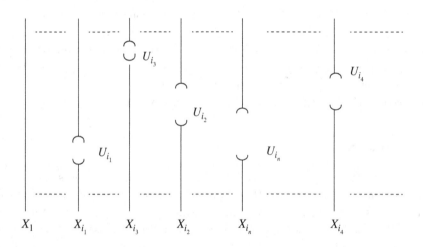

consists of all functions (tuples, if you wish) which assign to $k \neq i_1, \ldots, i_n$, any point of $X_k$, but must assign a point of $U_{i_1}$, to $i_1, \ldots$, and a point of $U_{i_n}$ to $i_n$.

**1. Definition.** Given the family $(X_i, \tau_i)$, $i = 1, 2, \ldots, n$, of topological spaces, the *product topology*, denoted by $\Pi\tau_i$, is the one which has $\Pi\mathcal{S}$ for a subbase.

The following is then obvious.

**2. Lemma.** The product topology $\Pi\tau_i$ on $\Pi_{i=1}^n X_i$ is the coarsest topology *with respect to* which all projections are continuous.

**3. Theorem.** A function $f : X \to \Pi_{i=1}^n X_i$ (precisely, $f : (X, \tau) \to (\Pi_{i=1}^n X_i, \Pi \tau_i)$) is continuous iff each $\Pi_j \circ f$ is continuous, for $j = 1, 2, \ldots, n$.

**Proof.** The "only if" part is immediate from Corollary 1.15.

    The "if" part: Note that, for each $U_j \in \tau_j$,

$$f^{-1}\Pi_j^{-1}(U_i) = (\Pi_j \circ f)^{-1}(U_j) \in \tau.$$

    Since $\Pi S = \{\Pi_j^{-1}(U_j) | U_j \in \tau_j, j = 1, 2, \ldots, n\}$ is a subbase for $\Pi \tau_i$, by Theorem 1.14(vi) we get that $f$ is continuous.

    The following two results, while trivial, seem to catch many by surprise. May they never surprise the reader again!

**4. Lemma.** For any spaces $X_1, \ldots, X_n$ and bijection $b : \{1, 2, \ldots, n\} \to \{1, 2, \ldots, n\}$,

$$\Pi_{i=1}^n X_i \simeq \Pi_{i=1}^n X_{b(i)}$$

(*i.e.*, the order of the factors $X_i$ in a product space is immaterial).

**Proof.** Simply define $\psi : (\Pi_{i=1}^n X_i) \to \Pi_{i=1}^n X_{b(i)}$, by letting

$$\psi(x_1, \ldots, x_n) = (x_{b(1)}, \ldots, x_{b(n)})$$

for each $(x_1, \ldots, x_n) \in \Pi_{i=1}^n X_i$. Then observe that

$$\psi(\Pi_j^{-1}(U_j)) = \Pi_{b(j)}^{-1}(U_{b(j)}), \quad \psi^{-1}(\Pi_{b(j)}^{-1}(U_{b(j)})) = \Pi_j^{-1}(U_j),$$

for subbase elements of $\Pi_{i=1}^n X_i$ and $\Pi_{i=1}^n X_{b(i)}$, respectively. Therefore $\psi$ and $\psi^{-1}$ are continuous, by Theorem 1.14 (vi), which completes the proof.

**5. Lemma.** For any finite collection $\{X_\alpha\}_{\alpha \in F}$ of spaces and partition $\mathfrak{p} = \{A_t | t \in \Lambda\}$ of $F$,

$$\Pi_{\alpha \in F} X_\alpha \cong \Pi_{t \in \Lambda}(\Pi_{\alpha \in \Lambda} X_\alpha)$$

(*i.e.*, introducing parenthesis at will in a product $A_1 \times A_2 \times \cdots \times A_m$ yields a homeomorphic product space).

**Proof.** Before proceeding with the proof, note that we are saying that, for example,

$$(X_1 \times X_2) \times X_3 \cong X_1 \times X_2 \times X_3 \cong (X_1 \times X_3) \times X_2.$$

    For the proof, simply let $\varphi : \Pi_{\alpha \in F} X_\alpha \to \Pi_{t \in \Lambda}(\Pi_{\alpha \in A_t} X_\alpha)$ be defined by letting $\varphi(t)$ be the element of $\Pi_{t \in \Lambda}(\Pi_{\alpha \in A_t} X_\alpha)$ such that

$$\Pi_\alpha(\Pi_t \varphi(f)) = f(\alpha),$$

for each $\alpha \in F$ (this may look horrible, but all it says is that $\varphi$ sends each tuple in $\Pi_{\alpha \in F} X_\alpha$ to the tuple of tuples in $\Pi_{\alpha \in A_i}$, $X_\alpha$, $t \in \Lambda$, with the same elements). The remainder of the proof is essentially the same as the proof of Lemma 4.

**An Application.** In Calculus, it is customary and convenient to think of a function $f : E^k \to E^m$ as an $m$-tuple $f = (f_1, \ldots, f_m)$ of real-valued functions such that

$$f(\bar{x}) = (f_1(\bar{x}), \ldots, f_m(\bar{x}))$$

for each $\bar{x} \in E^k$. It is generally hinted that indeed the $f_i$ are *functions* and that indeed $f$ *is continuous* iff *each* $f_i$, $i = 1, 2, \ldots, m$, *is continuous*. The riddle can be easily solved: Given $f : E^k \to E^m$ define $f_i : E^k \to E^1$ by $f_i = \Pi_i \circ f$, with $\Pi_i$ being the $i^{th}$-projection map. Then, by Theorem 3, $f$ is continuous iff each $f_i = \Pi_i \circ f$ is continuous.

The preceding application suggests the following constructions, which will find extensive use later on.

**6. Definition.** Given a space $X$, finite families $\{Y_i | i = 1, \ldots, n\}$, $\{Z_i | i = 1, \ldots, n\}$ of spaces and functions $f_i : X \to Y_i$, $g_i : Y_i \to Z_i$, $i = 1, \ldots, n$, let the functions (!)

$$(f_1, \ldots, f_n) : X \to \Pi_{i=1}^n Y_i, g_1 \times \cdots \times g_n : \Pi_{i=1}^n Y_i \to \Pi_{i=1}^n Z_i$$

be defined by

$$(f_1, \ldots, f_n)(x) = (f_1(x), \ldots, f_n(x)), \ g_1 \times \cdots \times g_n(y_1, \ldots, y_n) = (g_1(y_1), \ldots, g_n(y_n)).$$

Because of Theorem 3, the following result is immediate.

**7. Lemma.** The functions $(f_1, \ldots, f_n)$ and $g_1 \times \cdots \times g_n$ of Definition 6 are continuous if the functions $f_1, \ldots, f_n$ and $g_1, \ldots, g_n$ are continuous. The function $g_1 \times \cdots \times g_n$ is open if the functions $g_1, \ldots, g_n$ are open.

At this stage, the reader may wonder: But it seems that the definition of the product topology could be applied to infinite products verbatim. Can it? *Of course, it can!* Furthermore, it is easy to see that all results of this section are valid for infinite products. Indeed, with the possible exception of Lemma 5, the proofs of the other results apply to infinite products verbatim. We set off this fact by the following proposition.

**8. Proposition.** Lemmas 2, 4 and 7 and Theorem 3 remain valid for any product spaces.

## 2.2  Product Metrics and Topologies

Let $(X_i, d_i)$, $i = 1, 2, \ldots, n$, be a finite family of metric spaces. By analogy with the standard definition of distance in the cartesian plane (*i.e.*, $d((x_1, x_2), (y_1, y_i)) =$

$\sqrt{(x_1 - y_1)^2 + (x_2 - y_2)^2})$, we define a function $d : (\Pi_{i=1}^n X_i) \times (\Pi_{i=1}^n X_i) \to E^1$ by

$$d((x_1, \ldots, x_n), (y_1, \ldots, y_n)) = \sqrt{\Sigma_{i=1}^n d_i(x_i, y_i)^2}.$$

This function $d$ will always be referred to as the product-metric.

**9. Proposition.** The product-metric $d$ is actually a metric on $\Pi_{i=1}^n X_i$.

**Proof.** Certainly, we only need to verify the triangle inequality. But note that

$$(\Sigma_{i=1}^n d_i(x_i, y_i)^2)^{1/2} \leq (\Sigma_{i=1}^n (d_i(x_i, z_i)^2 + d_i(z_i, y_i)^2))^{1/2}$$
$$\leq (\Sigma_{i=1}^n d_i(x_i, z_i)^2)^{1/2} + (\Sigma_{i=1}^n d_i(z_i, y_i)^2)^{1/2},$$

the last inequality being a consequence of Minkowski's Inequality (see proof in Appendix A). This proves the triangle inequality for $d$.

**Remark.** The metric $\Gamma_1 : E^1 \times E^1 \to E^1$, defined by $\Gamma_1(x, y) = |x - y|$, and the resultant product-metrics on $E^n$, for $n = 2, 3, \ldots$, will be referred to as the *Euclidean metrics*. (Note that, in $E^n$, $\Gamma_n(\bar{x}, \bar{y}) = |\bar{x} - \bar{y}|$.)

**10. Proposition.** The topology generated by the product-metric $d$ on $\Pi_{i=1}^n X_i$ equals the product of the topologies generated by the metrics $d_i$ on $X_i$. That is, $\tau_d = \Pi \tau_{d_i}$.

**Proof.** Keeping Lemma 1.20 in mind, we only need to observe that

(a) For each $(x_1, \ldots, x_n) \in \Pi_{i=1}^n X_i$ and $\varepsilon > 0$,
$$B((x_1, \ldots, x_n), \varepsilon) \supset \Pi_{i=1}^n B(x_i, \varepsilon/\sqrt{n}).$$

(b) For each $(x_1, \ldots, x_n) \in \Pi_{i=1}^n X_i$, and $\varepsilon_i > 0$ for $i = l, \ldots, n$,
$$\Pi_{i=1}^n B(x_i, \varepsilon_i) \supset B((x_1, \ldots, x_n), \varepsilon),$$

where $\varepsilon = \min\{\varepsilon_1, \ldots, \varepsilon_n\}$.

Note that when we apply Proposition 10 to Euclidean spaces $E^n$, we then get that the topology generated by the collection of $n$-Euclidean balls

$$B((x_1, \ldots, x_n), \varepsilon) = \{(y_1, \ldots, y_n) \in E^n | (\Sigma_{i=1}^n |x_i - y_i|^2)^{1/2} < \varepsilon\})$$

equals the topology generated by the collection of open $n$-Euclidean cubes

$$\Pi_{i=1}^n ]a_i, b_i[ = \Pi_{i=1}^n B((a_i + b_i)/2, (b_i - a_i)/2).$$

If we did not have the benefit of Euclidean Geometry and of Pythagora's Theorem, it is only too possible that we would have chosen to distinguish another of the many metrics on cartesian products which are much easier to handle. For example, given the metric spaces $(X_i, d_i)$, $i = 1, \ldots, n$, it is trivial to check that the function $\rho : (\Pi_{i=1}^n X_i) \times (\Pi_{i=1}^n X_i) \to E^1$, defined by

$$\rho((x_1, \ldots, x_n), (y_1, \ldots, y_n)) = \Sigma_{i=1}^n d_i(x_i, y_i),$$

is a metric on $\Pi_{i=1}^n X_i$. It is easy to see that $\tau_d = \tau_\rho$. Yet $\rho$ offers many sobering surprises, when applied to the plane $E^2$:

(a) Using the standard techniques of calculus to measure the length of an arc we see that, with respect to $\rho$, the length of the hypotenuse of the triangle with vertices $(0, 0)$, $(1, 0)$, and $(1, 1)$ equals the sum of the lengths of the legs of the triangle (therefore, it is ambiguous to say that the *shortest distance between two points is a straight line*—it depends on the metric used).

(b) The ball $B((0,0),1)$, with respect to $\rho$ is not a "round" disc but a "square" one. Indeed, $B((0,0),1)$, is the square with corners $(1, 0)$, $(0, 1)$, $(-1, 0)$ and $(0, -1)$. Furthermore, using the standard horizontal-vertical grating techniques of calculus to find areas of surfaces, we would get that the area of the "square" disc $B((0,0),1)$ is 2.

## 2.3   Quotient Spaces

Given a topological space $(X,\tau)$, a *set* $Y$ and an onto function $f : X \twoheadrightarrow Y$ one is naturally (?) compelled to ask: Is there a "nice" topology $\sigma$ for $Y$ with respect to which $f$ is continuous? Certainly $f$ is continuous with respect to the indiscrete topology on $Y$; this can only tempt one to ask a more interesting question: *Is there a finest topology for $Y$ with respect to which $f$ is continuous?* If one exists, it certainly must be contained in

$$\tau_f = \{U \subset Y | f^{-1}(U) \in \tau\}.$$

But $\tau_f$ is a topology for $Y$, because of 0.15 (iv).

Therefore, $\tau_f$ is the topology we are looking for (since $f$ is not continuous with respect to any topology on $Y$ which is strictly finer than $\tau_f$).

**11.   Definition.** Let $(X,\tau)$ be a topological space, $Y$ a set and $f : X \twoheadrightarrow Y$ a function. The topology $\tau_f$ is called the *quotient topology* on $Y$, and $Y$ is called a *quotient space* of $X$ (with respect to $\tau$ and $f$). To say that $f : X \twoheadrightarrow Y$ is a *quotient function* means that the topology of $Y$ is the quotient topology (with respect to $f$ and the topology of $X$).

One of the most useful results of Topology is undoubtedly the following (in some ways, it is a dual of Theorem 3).

**12.   Theorem.** Let $f : X \twoheadrightarrow Y$ be a quotient function. Then a function $g : Y \to Z$ is continuous iff $g \circ f$ is continuous.

**Proof.** The "only if" part is obvious.

The "if" part: For each open $U \subset Z$, $(g \circ f)^{-1}(U) = f^{-1}(g^{-1}(U))$ is open in $X$. Since $f$ is a quotient function, we then get that, for each open $U \subset Z$, $g^{-1}(U)$ is open in $Y$, which shows that $g$ is continuous.

One of the major tasks in the applications of topological spaces is the construction of quotient spaces. It is therefore imperative that we consider the technique of constructing quotient spaces in its many guises:

*The Partition Technique* (really, the only way to go): Let $(X, \tau)$ be a topological space and $\mathcal{X} = \{A_\alpha\}_{\alpha \in \Lambda}$ a partition of $X$. Let $f : X \to \mathcal{X}$ be defined by "$f(x) =$ the $A_\alpha$ which contains $x$". Then

$$\{\mathcal{U} \subset \mathcal{X} | f^{-1}(\mathcal{U}) \in \tau\} = \tau_f = \{\mathcal{U} \subset \mathcal{X} | \bigcup \mathcal{U} \in \tau\}$$

is the quotient topology for $\mathcal{X}$.

*The Equivalence Relation Technique*: Let $(X, \tau)$ be a topological space and $R$ an equivalence relation on $X$. Let $X/R$ be the set of $R$-equivalence classes of $X$ (therefore, $X/R$ is a partition of $X$) and let $f : X \to X/R$ be defined by $f(x) =$ *the R-equivalence class* of $x$. Then

$$\{\mathcal{U} \subset X/R : f^{-1}(\mathcal{U}) \in \tau\} = \tau_f = \{\mathcal{U} \subset X/R : \bigcup \mathcal{U} \in \tau\}$$

is the quotient topology for $X/R$. (This technique is very useful whenever $X$ is rich in *natural* equivalence relations—for example, whenever $X$ supports group or ring operations.)

*The Adjunction (or "Gluing") Technique*: Let $(X, \tau)$, $(Y, \mu)$ be *disjoint* (i.e., $X \cap Y = \emptyset$) topological spaces, $A$ a closed subset of $X$ and $p : A \to p(A) \subset Y$ a continuous function. Let

$$\mathcal{X} = \{x | x \in X - A\} \cup \{\{y\} \cup p^{-1}(y) | y \in p(A)\} \cup \{y | y \in Y - p(A)\}.$$

Clearly $\mathcal{X}$ is a partition of $X \cup Y$; $\mathcal{X}$ is generally denoted by $X \bigcup_p Y$. Now that we are back to the partition technique, we proceed with the definition of $f : X \cup Y \to X \bigcup_p Y$ and the quotient topology $\tau_f$ in the expected manner. (Note that the restriction that $X \cap Y = \emptyset$ is somewhat artificial inasmuch that we can always give $X$ and $Y$ different colors; say $X \cong X \times \{0\}$ and $Y \cong Y \times \{l\}$.)

*The Identification Technique*: Let $f : (X, \tau) \to (Y, \mu)$ be an onto function. Then $X/f = \{f^{-1}(y) | y \in Y\}$ is a partition of $X$. Define $\bar{f} : X \to X/f$ by $\bar{f}(x) = f^{-1}f(x)$ ($\bar{f}$ is well defined!). Then

$$\{U \subset X/f : \bar{f}^{-1}(U) \in \tau\} = \tau_{\bar{f}} = \{U \subset X/f : \bigcup U \in \tau\}$$

is the quotient topology for $X/f$.

**13. Theorem.** Let $f : (X, \tau) \twoheadrightarrow (Y, \mu)$ be a quotient function. Then $X/f \cong Y$.

**Proof.** Define $h : Y \to X/f$ by $h(y) = f^{-1}(y)$. It is straightforward that $h$ is a

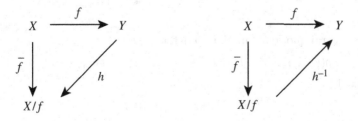

well-defined one-to-one and onto function, and that the diagrams are *commutative* (*i.e.*, $h \circ f = \bar{f}, \ldots$). Therefore, by Theorem 12, both $h$ and $h^{-1}$ are continuous.

It is clear from Corollary 1.16 that all homeomorphisms are quotient functions. Indeed, we have the much stronger result.

**14. Lemma.** If $f : (X, \tau) \twoheadrightarrow (Y, \mu)$ is continuous and open (or closed) then $f$ is a quotient function.

**Proof.** Without loss of generality, let us assume that $f$ is closed. Note that, if $f^{-1}(U) \in \tau$ then $U = Y - f(X - f^{-1}(U)) \in \mu$, since $f$ is closed. Therefore, $\tau_f \subset \mu$. Since $f$ is continuous, we clearly have that $\mu = \tau_f$. Therefore, $\mu = \tau_f$ and $f$ is a quotient function.

## 2.4   Applications

*Cones.* For any space $X$, let $\hat{N} = \{(x, 1) | x \in X\}$ and let
$$CX = \{\hat{N}\} \cup \{(x, t)\} | x \in X, 0 \le t < 1\}.$$
$CX$, with the quotient topology with respect to the function $f : X \times I \to CX$ (defined by $f((x, 1)) = \hat{N}$ and $f((x, t)) = \{(x, t)\}$ for $t < 1$), where $X \times I$ has the product topology, is called the *cone of* $X$. (Note that, if $X = S^1$, then $S^1 \times I$ is an "ordinary" cylinder and $CS^1$ is really obtained from $S^1 \times I$ by "pinching" the top to a point—that is, $CS^1$ is really (homeomorphic to) a cone, which is really a closed disc.)

**15. Proposition.** $CS^n \simeq B^{n+1}$, for $n = 0, 1, \ldots$.

**Proof.** Let us consider the diagram

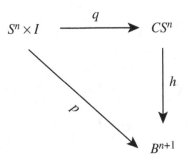

with $q$ the natural quotient function, $p$ defined by
$$p((x)_o, \ldots, x_n), t) = ((1 - t)x_o, \ldots, (1 - t)x_n) \in B^{n+1}$$
and $h$ defined by
$$h(\hat{N}) = (0, 0, \ldots, 0) \in B^{n+1},$$

$$h(\{((x_o, \ldots, x_n), t)\}) = ((1 - t)x_o, \ldots, (1 - t)x_n).$$

(Are $p$ and $h$ well defined? Is $h$ one-to-one?) At this stage, the reader may have considerable difficulty in proving that $p$ is a quotient function; nonetheless, this is immediate from Theorem 3.7 and Lemma 14. From Theorem 12, we then get that $h$ and $h^{-1}$ are continuous.

*Suspensions.* For any space $X$, let $\hat{N} = \{(x, 1)|x \in X\}$ and $\hat{S} = \{(x, -1)|x \in X\}$. Then let

$$SX = \{\hat{N}, \hat{S}\} \cup \{(x, t)\}|x \in X, -1 < t < 1\},$$

with the quotient topology with respect to the function $f : X \times [-1, 1] \to SX$, defined by $f(x, 1) = \hat{N}$, $f(x, -1) = \hat{S}$, and $f(x, t) = \{(x, t)\}$, for $-1 < t < 1$, is called the *suspension* of $X$. (Note that, if $X = S^1$, then $X \times [-1, 1]$ is a cylinder and $SS^1$ is really obtained from $X \times [-1, 1]$ by "pinching" the top rim to a point and *pinching* the bottom rim to another point—that is, $SS^1$ is really $S^2$. $\hat{N}$ becomes the *north* pole and $\hat{S}$ becomes the *south* pole of $S^2$.)

**16. Proposition.** $SS^n = S^{n+1}$ for $n = 0, 1, \ldots$.

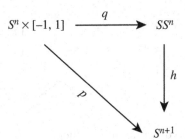

**Proof.** Let us consider the diagram above with $q$ the natural quotient map, $p$ defined by

$$p((x_o, \ldots, x_n), t) = \begin{cases} ((1 - t)x_o, \ldots, (1 - t)x_n, \ \sqrt{1 - (1 - t)^2}), & \text{for } t \geq 0 \\ ((1 + t)x_o, \ldots, (1 - t)x_n, \ -\sqrt{1 - (1 + t)^2}), & \text{for } t \leq 0 \end{cases}$$

and $h$ defined by

$$h(N) = (0, \ldots, 0, 1), h(S) = (0, \ldots, 0, -1) \text{ and}$$

$$h((x_o, \ldots, x_n), t) = \begin{cases} ((1 - t)x_o, \ldots, (1 - t)x_n, \ \sqrt{1 - (1 - t)^2}), & \text{for } t \geq 0 \\ ((1 + t)x_o, \ldots, (1 - t)x_n, \ \sqrt{1 - (1 + t)^2}), & \text{for } t \leq 0. \end{cases}$$

Again the reader may find it difficult to prove that $p$ is a quotient function; it is nonetheless immediate from Theorem 3.7 and Lemma 14. From Theorem 12, we then get that $h$ and $h^{-1}$ are continuous.

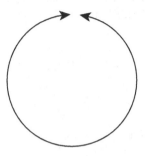

*Boundary Identifications.* The following result justifies the old-fashioned technique of making circles—one simply "glues" or "identifies" the endpoints of a line segment.

**17. Proposition.** $B^n/S^{n-1} \simeq S^n$, for $n = 1, 2, \ldots$. (See 0.16.)

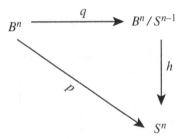

**Proof.** Let us consider the diagram above where $q$ is the natural quotient map and $p$ is defined by (recall that $\bar{x} = (x_1, \ldots, x_n)$ and $|\bar{x}| = \sqrt{x_1^2 + \cdots + x_n^2}$)

$$p(\bar{x}) = \begin{cases} (4(1 - |\bar{x}|)x_1, \ldots, 4(1 - |\bar{x}|)x_n, \ \sqrt{1 - 16(1 - |\bar{x}|^2)}), & \text{for } |\bar{x}| \geq 1/2, \\ (4|\bar{x}|x_1, \ldots, 4|\bar{x}|x_n, \ -\sqrt{1 - 16x^4)}), & \text{for } |\bar{x}| \leq 1/2, \end{cases}$$

($p$ is well defined!) and $h$ is defined by $h(S^{n-1}) = (0, \ldots, 0, 1)$, and $h([\bar{x}]) = p(\bar{x})$ whenever $|\bar{x}| < 1$ ($h$ is well defined and 1–1!).

Again, we get that $p$ is a quotient map, because of Lemma 14 and Theorem 3.7. By Theorem 12, we then get that $h$ and $h^{-1}$ are continuous.

By now the reader must be discouraged by the annoying fact that the precise construction of even the elementary quotient spaces just described presents painful details. Fortunately, topologists have devised descriptive techniques of construction of quotient spaces—commonly called the *Scissors-and-Paste Techniques*. The following three examples illustrate these techniques.

*The Möbius Band.* The Möbius Band is obtained from *the square* $I \times I$ by *gluing* each $(0, t)$ with each $(1, 1 - t)$, for $0 \le t \le 1$. Descriptively,

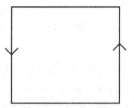

which successfully yields the graphic representations

*The Torus.* The Torus is obtained from the square $I \times I$ by *gluing* each $(0, t)$ with each $(1, t)$ and also *gluing* each $(t, 0)$ with each $(t, 1)$, for $\le t \le 1$. Descriptively,

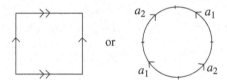

which successfully yields the graphic representations

*The Klein Bottle.* Descriptively, the Klein Bottle is

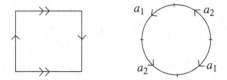

No one has ever really seen a graphic representation of a Klein Bottle. Certainly the first step of identifying the upper and lower edges of the square produces a cylinder.

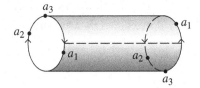

One can immediately see that a 180°-twist in the middle of the cylinder will simply not match the points $a_1$, $a_2$, $a_3$ of the left-rim with $a_1$, $a_2$, $a_3$ of the right-rim, respectively. Indeed, in $E^3$ nothing works (this is quite difficult to prove!). Nonetheless, the following *erroneous* graphic representation of the Klein Bottle in $E^3$ helps one visualize what it looks like in $E^n$ with $n \geq 4$.

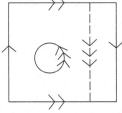

The trouble is that one cannot *cut* through the side of the cylinder—that is really an identification which is not called for.

Indeed, the reader should note that the preceding picture is the graphic representation of

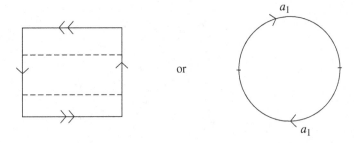

*The Projective Plane.* Descriptively, the Projective Plane is

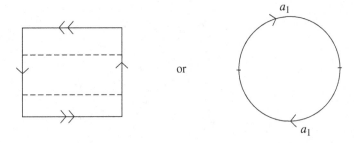

or

and, therefore, it is obtained from a closed disc by *gluing* antipodal points (*i.e.*, diametrically opposed points) of its boundary. It is difficult to give a graphic representation of the projective plane. Nonetheless, if we cut the projective plane along the dotted lines, we get the

Möbius Band

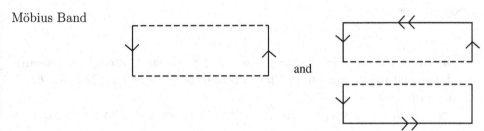

and

But, by successive *gluings*, we see that

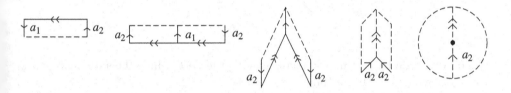

Therefore, the projective plane is obtained from the Möbius Band and a closed disc by *gluing* their edges together.

## Chapter 2. Exercises

1. Show that the function $f : X \twoheadrightarrow Y$ is a quotient function iff the following condition holds: $B \subset Y$ is closed iff $f^{-1}(B)$ is closed in $X$.

2. Show that open (or closed) continuous functions are quotient functions.

3. Let $p : X \twoheadrightarrow Y$ be a quotient function and $f : X \to Z$ be a continuous function such that $f \circ p^{-1} : Y \to Z$ is a function (note that $f \circ p^{-1}$ is a function iff each $f^{-1}(z) \subset$ some $p^{-1}(y)$). Show $f \circ p^{-1}$ is continuous.

4. Show that the function $f : (X, \tau) \twoheadrightarrow (Y, \mu)$ is an *open* function iff the following condition holds: There exists a base $\mathcal{B}$ for $\tau$ such that $f(B) \in \mu$ for each $B \in \mathcal{B}$. (Hint: use 0.15(vi)).

5. Show that all projections of the product space $(\Pi_{i=1}^{n} X_i, \Pi \tau_i)$ are *open* functions. (Hint: see ex. 4.)

6. Show that the torus is (homeomorphic to) $S^1 \times S^1$. (Indeed, it is not uncommon to define the torus as the product space $S^1 \times S^1$.)

7. Show that the torus is really a 2-*sphere* $S^2$ *with a tubular handle*.

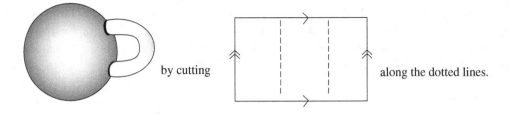

by cutting          along the dotted lines.

8. Show that the torus is a quotient space of $E^2$. (Hint: Grate $E^2$ into squares of base and height equal to 1. Then describe the quotient function from $E^2$ to the torus in descriptive form.)

9. Try to figure out a graphic representation of the quotient space, descriptively

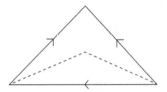

of a triangle by cutting the triangle along the dotted line. This quotient space is known as the Dunce Hat.

10. Show that the quotient space of the square, descriptively

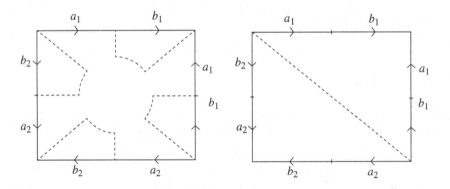

is a 2-sphere with two handles, by cutting along the dotted lines of the square on the left. However, by cutting along the dotted line of the square on the right, show that the same quotient space of the square really is a double-torus (two torii glued along the edges of a hole on each of them).

11. Show that a 2-sphere with two handles and a double-torus are homeophoric (see ex. 10).

12. Let $X \xrightarrow{f} Y \xrightarrow{g} Z$ such that $f$ and $g$ are continuous and $g \circ f$ is a quotient function. Show that $g$ is a quotient function.

13. Let $f : X \twoheadrightarrow Y$ be a quotient function. Show that if $f$ is 1–1, then $f$ is a homeomorphism.

14. For each $\alpha \in \Lambda$, let $p_\alpha : X_\alpha \to Y_\alpha$ be a continuous open onto map. Show that $\Pi p_\alpha : \Pi_\alpha X_\alpha \to \Pi_\alpha Y_\alpha$ is a quotient map.

15. Show that the composition of two quotient maps is a quotient map.

16. Let $R_1$ and $R_2$ be two equivalence relations in a space $X$ such that $x\,R_1\,y$ implies $x\,R_2\,y$, for all $x$, $y \in X$. Show that $X/R_1$ is a quotient space of $X/R_2$.

17. Let $(X, \tau)$ be a space and $A \subset X$ a subspace. Assume there exists a continuous $r : X \to A$ such that $r(a) = a$, for each $a \in A$ ($r$ is called a *retraction*). Show that $r$ is a quotient map.

18. Let $\{(X_\alpha, \tau_\alpha)\}_{\alpha \in \Lambda}$ be any family of spaces; also let $X = \Pi_{\alpha \in \Lambda} X_\alpha$, $\tau = \Pi \tau_\alpha$ and $\Pi_\alpha : X \to X_\alpha$ be the $\alpha$-projection for each $\alpha \in \Lambda$ (cf. 0.18). Also, for each $f \in X$ and $\Gamma \subset \Lambda$, let

$$S(f, \Gamma) = \{g \in X | g(\alpha) = f(\alpha), \text{ for all } \alpha \in \Lambda - \Gamma\}.$$

The subspace $S(f, \Gamma)$ of $X$ will be referred to as the $(f, \Gamma)$-*slice of $X$*.

   (a) Prove that $S(f, \Gamma) \cong \Pi_{\alpha \in \Gamma} X_\alpha$. (Hint: Define $h : S(f, \Gamma) \to \Pi_{\alpha \in \Gamma} X_\alpha$ by $[h(g)](\alpha) = g(\alpha)$, for every $\alpha \in \Gamma$. Show $h$ is 1–1 and onto. Think of $h^{-1} : \Pi_{\alpha \in \Gamma} X_\alpha \to \Pi_{\alpha \in \Lambda} X_\alpha$. Use Theorem 3 and Remark 4 to show that $h$ and $h^{-1}$ are continuous.)

   (b) Prove that, for each $g \in X$ and $\{\beta_1, \ldots, \beta_n\} \subset \Lambda$, $S(g, \{\beta_1, \ldots, \beta_n\}) \simeq \Pi_{i=1}^n X_{\beta_i}$.

19. Let $X$ be a set and $\mathcal{A} = \{(A_\alpha, \tau_\alpha) | \alpha \in \Lambda\}$ be a family of spaces such that $\bigcup_\alpha A_\alpha = X$. Assume that

   (i) $\tau_\alpha | A_\alpha \cap A_\beta = \tau_\beta | A_\alpha \cap A_\alpha$. For all $\alpha$, $\beta \in \Lambda$
   (ii) each $A_\alpha \cap A_\beta$ is closed in $A_\alpha$, and in $A_\beta$.

Let $\tau(\mathcal{A}) = \{U \subset X | U \cap A_\alpha \in \tau_\alpha\}$. Show that

   (a) $\tau(\mathcal{A})$ is a topology for $X$ (called, *the weak topology over* $\mathcal{A}$)
   (b) $\tau(\mathcal{A}) | A_\alpha = \tau_\alpha$, for every $\alpha \in \Lambda$.

20. Given a set $X$, a space $(Y, \tau)$ and a function $f : X \twoheadrightarrow Y$, show that

   (a) $\tau^f = \{f^{-1}(U) | U \in \tau\}$ is a topology on $X$.
   (b) $f : (X, \tau^f) \twoheadrightarrow (Y, \tau)$ is a quotient map.

21. Given a set $X$, a collection $\{(Y_\alpha, \tau_\alpha) | \alpha \in \Lambda\}$ spaces and functions $f_\alpha : X \to Y_\alpha$, for each $\alpha \in \Lambda$, let $\tau$ be the topology generated by $\mathcal{S} = \{f_\alpha^{-1}(U) | U \in \tau_\alpha, \alpha \in \Lambda\}$. Show that

   (a) If $j : X \to \Pi_{\alpha \in \Lambda} Y_\alpha$ is defined by $(j(x))_\alpha = f_\alpha(x)$, for each $\alpha \in \Lambda$, then $\tau = \tau^j$ (see preceding problem).
   (b) If $x \neq w$ implies that there exists $\alpha \in \Lambda$ such that $f_\alpha(x) \neq f_\alpha(w)$, show that $j$ is a homeomorphism (between $X$ and $j(X)$).

22. Let $Q$ be the space of rational numbers, $R$ the identity relation on $Q$, $S$ the relation on $Q$ which identifies all the integers. Show that $(Q \times Q)/(R \times S)$ is not homeomorphic to $(Q/R) \times (Q/S)$. In particular, note that even though the natural maps $p : Q \to Q/R$ and $q : Q \to Q/S$ are quotient maps, their product $p \times q : Q \times Q \to Q/R \times Q/S$ is not a quotient map.

# Chapter 3

# Very Special Spaces

While studying calculus, the reader must have become well aware of the importance of the *compact* (*i.e.* closed and bounded) subsets of the real line, because of the results: The continuous image of a compact subset of $E^1$ is compact. The continuous image of a closed interval is a closed interval (equivalently, for each continuous function $f : [a, b] \to E^1$ and $f(a) \le d \le f(b)$ there exists $a \le e \le b$ such that $f(e) = d$ (the Intermediate-Value Theorem!)). Every continuous real-valued function on a closed interval of $E^1$ attains a maximum and a minimum value. Every Cauchy sequence $\{x_n\}$ (*i.e.*, for every $\varepsilon > 0$ there exists integer $n(\varepsilon)$ such that $m$, $n > n(\varepsilon)$ implies $|x_m - x_n| < \varepsilon$) of real numbers converges to some real number. Every sequence in a compact subset $A$ of $E^1$ has a subsequence which converges in $A$.

Many more equally important results are easily obtained once one truly understands the concepts just mentioned (more precisely, the mathematical concepts just alluded to). We are now ready for those *very special spaces*: Compact spaces, complete metric spaces, connected spaces and arcwise connected spaces.

But first we need more terminology, including a tiny bit of the hierarchy of topological spaces.

A. $T_1$, $T_2$, $T_3$: Given that topological spaces are not always metrizable (cf. ex 1.28) it is only natural to classify topological spaces in accordance with how similarly they behave like metric spaces. For this reason, topological spaces have been divided into many classes (dozens), there being no universal agreement on these divisions. Let $(X, \tau)$ be a topological space. Then

(i) $(X, \tau)$ is a $T_1$-*space* provided that $X - \{p\} \in \tau$ for each $p \in X$ (equivalently, for any $x$, $y \in X$ with $x \neq y$, there exists neighborhoods $N_x$ of $x$ and $N_y$ of $y$ such that $y \notin N_x$ and $x \notin N_y$).

(ii) $(X, \tau)$ is a $T_2$-*space* or a *Hausdorff space* provided that for any $x$, $y \in X$, with $x \neq y$, there exists neighborhoods $N_x$ of $x$ and $N_y$ of $y$ such that $N_x \cap N_y = \emptyset$.

(iii) $(X, \tau)$ is a *regular space* or a $T_3$-*space* provided that $(X, \tau)$ is $T_1$ and, for any $x \in X$ and closed set $B \subset X - \{x\}$, there exist neighborhoods $N_x$ of $x$ and $N_B$ of $B$ such that $N_x \cap N_B = \emptyset$; equivalently, for any $x \in X$ and neighborhood $U$ of $x$ there exists a neighborhood $V$ of $x$ such that $V^- \subset U$.

(Some mathematicians do not require that $X$ be $T_1$ in the definition of a regular space and then they let

$$T_3\text{-}space \equiv \text{regular space which is also } T_1.$$

This story has two morals—the first is that whenever one reads a scientific book one should make sure of the terminology used in that book; the second is that intellectuals sometimes assert their intellectual freedom in the most childish way.)

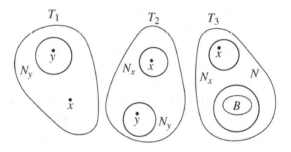

The following proposition should be obvious.

**1. Proposition.** $X$ is metrizable implies $X$ is $T_3$ implies $X$ is $T_2$ implies $X$ is $T_1$.

**Proof.** Only the first implication may not be absolutely trivial, but even that one is a straightforward consequence of Theorem 1.8(iv).

B. *Accumulation, Clustering, Converging.* Let $(X, \tau)$ be a topological space. Then

(i) For $A \subset X$, a point $p \in X$ is said to be an *accumulation point of $A$* or *cluster point of $A$*, provided that, for each neighborhood $N_p$ of $p$, $N_p \cap (A - \{p\}) \neq \emptyset$.

(ii) A sequence $\{x_n\}$ in $X$ is said to *cluster at the point $p \in X$* (*or $p$ is a cluster point of $\{x_n\}$*) provided that either $\{x_n\}$ is infinite and $p$ is a cluster point of the set $\{x_n | n \in \mathbf{N}\}$ or $\{x_n\}$ is finite and $p = x_j = x_{j+1} = \cdots$ for some $j$. (Be assured that *this unusual definition has its merits*.)

(iii) A sequence $\{x_n\}$ in $X$ is said to *converge to the point $p \in X$*, provided that, for each neighborhood $N_p$ of $p$, there exists some integer $n(N_p)$ such that $\{x_j | j \geq n(N_p)\} \subset N_p$; we let

$$\lim_n x_n = p \equiv \{x_n\} \text{ converges to } p.$$

**2. Proposition.** The following statements are valid in any space $X$:

    (a) if $\lim\limits_{n} = x$ and $\{w_n\}$ is a subsequence of $\{x_n\}$ then $\lim\limits_{n} w_n = x$.

    (b) A point $p \in X$ is a cluster point of $A \subset X$ iff $p \in (A - \{p\})^-$.

**Proof.** We only prove part (b), since the proof of (a) is too trivial.

The "if" part: Assume there exists some open neighborhood $N_p$ of $p$ such that $N_p \cap A = \emptyset$. Then $X - N_p$ is closed, $p \notin X - N_p$, and $A - \{p\} \subset X - N_p$, contradicting the hypothesis that $p \in (A - \{p\})^-$.

The "only if" part: Assume $p \notin (A - \{p\})^-$. Then there exists a closed set $B$ such that $A - \{p\} \subset B$ and $p \notin B$. Then $X - B$ is a neighborhood of $p$ such that $(X - B) \cap (A - \{p\}) = \emptyset$, a contradiction.

**3. Lemma.** Let $X$ be Hausdorff and $\{x_n\}$ a sequence in $X$. If $\lim\limits_{n} x_n = z$ and $\lim\limits_{n} x_n = w$ then $z = w$.

**Proof.** Suppose $w \neq z$. Pick disjoint open neighborhoods $N_w$, $N_z$ of $w$ and $z$, respectively. Then there exist integers $n(N_w)$ and $n(N_z)$ such that

$$j > n(N_w) \text{ implies } x_j \in N_w \text{ and } k > n(N_z) \text{ implies } x_k \in N_z.$$

Pick $t > \max\{n(N_w), n(N_z)\}$. Then $x_t \in N_w \cap N_z$, a contradiction.

## 3.1 Compact Spaces

Certainly it is ridiculous to expect that we define a compact space $A$ as a bounded and closed subset of a topological space $(X, \tau)$, since the notion of "bounded set" makes little sense in a space without a notion of distance. The surprise is that this same definition would be equally ridiculous in a metric space $(X, d)$—for example, let $X = E^1 - \{0\}$ and $d$ be the Euclidean metric on $X$; then $A = [-1, 1] \cap X$ is a closed and bounded subset of $X$ and yet not every sequence in $A$ has a subsequence which converges in $A$ (try $\{\frac{1}{n}\}$). There should be no doubt that the definition of compact subsets of $E^1$ depends on a very peculiar combination of circumstances.

**4. Definition.** A topological space $(X, \tau)$ is *compact* provided that each open cover $\mathcal{U}$ of $X$ (*i.e.*, $\bigcup \mathcal{U} = X$ and $\mathcal{U} \subset \tau$) contains a finite subcover $\mathcal{V}$ of $X$ (*i.e.*, $\bigcup \mathcal{V} = X$, $\mathcal{V}$ is finite). A metric space $(X, d)$ is compact if $(X, \tau_d)$ is compact.

**5. Lemma.** Let $(X, \tau)$ be a topological space and $\mathcal{B}$ any base for the topology $\tau$. If each open cover $\mathcal{U} \subset \mathcal{B}$ of $X$ has a finite subcover then $(X, \tau)$ is compact.

**Proof.** Let $\mathcal{U}$ be *any* open cover of $X$. By the definition of a base for a topology, there exists another open cover $\mathcal{V} \subset \mathcal{B}$ such that, for each $V \in \mathcal{V}$, $V \subset$ some $U \in \mathcal{U}$. By hypothesis, there exists a finite subcover $\{V_1, \ldots, V_n\}$ of $\mathcal{V}$. Pick any $U(V_i) \in \mathcal{U}$

such that $V_i \subset U(V_i)$ for $i = 1, \ldots, n$. Then $\{U(V_1), \ldots, U(V_n)\}$ is a finite subcover of $\mathcal{U}$.

**6. Lemma.** The following are valid:

(a) A closed subset of a compact space is compact.
(b) A compact subspace of a Hausdorff space is closed.
(c) The continuous image of a compact space is compact.

**Proof.**

(a) Let $(X, \tau)$ compact and $A$ a closed subspace of $X$. If $\mathcal{U}$ is an open cover of $A$ then $\mathcal{U}' = \{U \cup (X - A) | U \in \mathcal{U}\}$ is an open (!) cover of $X$. Pick finite subcover $\{U_1 \cup (X - A), \ldots, U_n \cup (X - A)\}$ of $\mathcal{U}'$. Then $\{U_1, \ldots, U_n\}$ is a finite subcover of $\mathcal{U}$.

(b) Suppose $(X, \tau)$ Hausdorff, $A$ a compact subspace of $X$ which is not closed. Let $p \in A^- - A$. Then, for each $a \in A$, there exist open neighborhoods $N_{pa}$ of $p$ and $N_a$ of $a$ such that $N_{pa} \cap N_a = \emptyset$. Then $\mathcal{U} = \{N_a | a \in A\}$ is an open cover of $A$ with no finite subcover. (Suppose $\{N_{a_1}, \ldots, N_{a_n}\}$ is a finite subcover of $\mathcal{U}$. Then $\bigcap_{i=1}^{n} N_{pa_i}$ is a neighborhood of $p$ which misses $A - (\bigcap_{i=1}^{n} N_{pa_i}) \cap (\bigcup_{i=1}^{n} N_{a_i}) = \emptyset$ — a contradiction.)

(c) Let $f : X \to Y$ be continuous and let $X$ be compact. If $\mathcal{U}$ is an open cover of $Y$, then $\mathcal{U}' = \{f^{-1}(U) | U \in \mathcal{U}\}$ is an open cover of $X$. If $\{f^{-1}(U_i) | i = 1, \ldots, n\}$ is a finite subcover of $\mathcal{U}'$ then $\{U_i | i = 1, \ldots, n\}$ is a finite subcover of $\mathcal{U}$.

**7. Theorem.** Let $f : X \to Y$ be continuous, with $X$ compact and $Y$ Hausdorff. Then, $f$ is a closed function and, therefore, a quotient function.

**Proof.** Let $A$ be a closed subset of $X$. Then, from Lemma 6(a), (c), (b) we get that $f(A)$ is closed. Therefore, by Lemma 2.7, $f$ is a quotient map.

**8. Theorem.** For any $n \in \mathbf{N}$, let $(X_1, \tau), \ldots, (X_n, \tau_n)$ be compact spaces. Then $(\Pi_{i=1}^{n} X_i, \Pi\tau_i)$ is compact.

**Proof.** (By induction.) Clearly $(X_1, \tau_1)$ is compact. Assume $(\Pi_{i=1}^{n-1} X_i, \Pi\tau_i)$ is compact and let us show that $(\Pi_{i=1}^{n} X_i, \Pi\tau_i)$ is compact: Because of Lemma 5 we consider only open covers $\mathcal{U}$ of $\Pi_{i=1}^{n} X_i$ whose elements have the form $\Pi_{i=1}^{n} U_i$ with $U_i \in \tau_i$, $i = 1, 2, \ldots, n$. Since, for each $\bar{x} = (x_1, \ldots, x_{n-1}) \in \Pi_{i=1}^{n-1} X_i$, $\{\bar{x}\} \times X_n \cong X_n$ and $\{U \cap (\{\bar{x}\} \times X_n) | U \in \mathcal{U}\}$ is an open cover of $\{\bar{x}\} \times X_n$, we can find $\Pi_{i=1}^{n} U_i^1, \ldots, \Pi_{i=1}^{n} U_i^k \in \mathcal{U}$ which cover $\{\bar{x}\} \times X_n$. Let $C_{\bar{x}} = (\bigcap_{j=1}^{k} \Pi_{i=1}^{n-1} U_i^j) \times (\bigcup_{j=1}^{k} U_n^j) = J_{\bar{x}} \times K_{\bar{x}}$, for each $\bar{x} \in \Pi_{i=1}^{n-1} X_i$. Here is a visual description of the preceding.

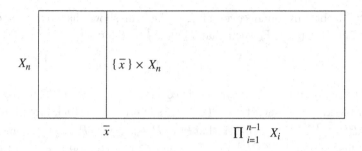

Since $\{J_{\bar{x}}|\bar{x} \in \Pi_{i=1}^{n-1}X_i\}$ is an open cover of the compact space $\Pi_{i=1}^{n-1}X_i$, we can find $J_{\bar{x}_i}, \ldots, J_{\bar{x}_m}$ which cover $\Pi_{i=1}^{n-1}X_i$. Therefore $\{C_{\bar{x}_1}, \ldots, C_{\bar{x}_m}\}$ covers $\Pi_{i=1}^{n}X_i$, which implies that *there is a finite number of elements of* $\mathcal{U}$ *which cover* $\Pi_{i=1}^{n}X_i$, since each $C_{\bar{x}_r}$ is covered by finitely many elements of $\mathcal{U}$, for $r = 1, 2, \ldots, m$.

**9. Theorem.** For any metric space $(X, d)$ and $C \subset X$, we get that (a) iff (b) iff (c) implies (d). If $(X, d) = (E^n, $ Euclidean metric$)$ then (a) iff (b) iff (c) iff (d).

   (a) $C$ is compact,
   (b) Every infinite subset $A$ of $C$ has a cluster point in $C$,
   (c) Every sequence $\{x_n\}$ in $C$ has a subsequence which converges in $C$,
   (d) $C$ is a closed and bounded subset of $(X, d)$.

**Proof.** First, (a) implies (b): Suppose $A \subset C$ is infinite with no cluster point. Pick sequence $\{x_n\}$ of distinct points of $A$. Then the set $B = \{x_n | n \in \mathbf{N}\}$ has no cluster point which implies that

  (i) $B$ is closed: If not, any point in $B^- - B$ would be a cluster point of $B$ (cf. Prop. 2),
  (ii) For each $n$, there exists an open neighborhood $U_n$ of $x_n$ such that $U_n \cap (B - \{x_n\}) = \emptyset$: If not, $x_n$ would be a cluster point of $B$. Then $\{C - B\} \cup (U_n | n \in \mathbf{N}\}$ is an open cover of $C$ with *no* finite subcover (each $U_n$ covers no $x_j$, with $j \neq n$, and $C - B$ covers no $x_j$ at all), a contradiction.

   (b) implies (c). Let $\{x_n\}$ be a sequence in $C$. Without loss of generality, we assume the range of $\{x_n\}$ is infinite. Then there exists a cluster point $p$ for the set $A = \{x_n | n \in \mathbf{N}\}$. Inductively, it is easy to pick $x_n \in A$ such that $d(x_{n_k}, p) < \frac{1}{k}$ and $n_k < n_{k+1}$ for $k \in \mathbf{N}$. It follows that $\{x_{n_k}\}$ is a subsequence of $\{x_n\}$ such that $\lim_{k} x_{n_k} = p$.

   (c) implies (a). First observe that, for each $\varepsilon > 0$, there exists a finite subset $F_\varepsilon$ of $C$ such that $C = \bigcup\{B(u, \varepsilon) | u \in F_\varepsilon\}$ ($F_\varepsilon$ is called an $\varepsilon$-net): Suppose this is not true for some $\varepsilon_0 > 0$. Then, by induction, one can immediately pick a sequence $\{x_n\}$ in $C$ such that $d(x_k, x_{n+1}) \geq \varepsilon_0$ for $k \leq n$, $n \in \mathbf{N}$. Clearly, $\{x_n\}$ has no convergent subsequence, a condiction.

Now let $\mathcal{U}$ be any open cover of $X$. We will show that $\mathcal{U}$ has a countable subcover: Let $D = \bigcup_{k=1}^{\infty} F_k$ such that $F_k$ is a $\frac{1}{k}$-*net* and let

$$\mathcal{D} = \left\{ B\left(d, \frac{1}{j}\right) \subset \text{ some } U \in \mathcal{U} | j \in \mathbf{N} \text{ and } d \in D \right\}.$$

Clearly, $\mathcal{D}$ is countable. Furthermore, $\mathcal{D}$ covers $C$. (Let $p \in C$. Then there exists some positive integer $j$ such that $B(p, \frac{1}{j}) \subset$ some $U_p \in \mathcal{U}$. Pick any $u_p \in F_{2j}$ such that $p \in B(u_p, \frac{1}{2j})$). We then get that $p \in B(u_p, \frac{1}{2j}) \subset B(p, \frac{1}{j}) \subset$ some $U_p \in \mathcal{U}$, which implies that $p \in B(u_p, \frac{1}{2j}) \in \mathcal{D}$.) Since $\mathcal{D}$ is a countable cover of $C$, by the definition of $\mathcal{D}$ one immediately gets that $\mathcal{U}$ has a countable subcover $\{U_n | n \in \mathbf{N}\}$.

Finally, we show that the countable cover $\{U_n | n \in \mathbf{N}\}$ of $C$ has a finite subcover: Suppose not. Pick $x_n \in C - \bigcup_{i=1}^{n} U_i$ for $n \in \mathbf{N}$. By hypothesis, $\{x_n\}$ has a convergent subsequence $\{x_{n_k}\}$; say $\lim_k X_{n_k} = q$ and $q \in U_m$. Then $U_m$ is a neighborhood of $q$ which misses $\{x_n | n \geq m\}$, which shows that $\{x_{n_k}\}$ does not converge to $q$, a contradiction. This does the trick.

We have, thus far, proved that (a) iff (b) iff (c). Much has been said in the proof of (c) implies (a). (See ex. 18.)

(c) implies (d). Suppose $C$ is not closed in $(X, d)$: Pick any $p \in C^- - C$. For each $n$, pick $x_n \in C \cap B(p, \frac{1}{n})$. Then $\lim_n x_n = p$. Therefore, for any subsequence $\{x_{n_k}\}$ of $\{x_n\}$, $\lim_k x_{n_k} = p$, contradicting (c).

Suppose $C$ is not bounded in $(X, d)$: Then it is easily seen that, for any $c \in C$,

$$\{B(c, n) \cap C | n \in \mathbf{N}\}$$

is an open cover of $C$ with no finite subcover (say $B(c, n_1) \cap C, \ldots, B(c, n_k) \cap C$ cover $C$, with $n_1 < \ldots < n_k$. Then $C \subset B(c, n_k)$, implying that $C$ is bounded).

Finally, we prove that, for Euclidean spaces, (d) implies (a). (Clearly, this yields (a) iff (b) iff (c) iff (d) in Euclidean space.) First (d) implies (a) in $E^1$: Without loss of generality, we let $C$ be some closed interval $[m, M]$ (because of Lemma 6(a)) and consider only open covers $\mathcal{O}$ of $[m, M]$ whose elements are "open intervals intersected with $[m, M]$" (because of Lemma 5 and the definition of the subspace topology). Indeed, note that, because of the definition of the subspace topology we can, just as well, let $\mathcal{O}$ be a cover of $[m, M]$ by open intervals of $E^1$, for then $\{0 \cap [m, M] | 0 \in \mathcal{O}\}$ is a cover of $[m, M]$ of the required form.

So let $\mathcal{O}$ be a cover of $[m, M]$ by open intervals in $E^1$. Let us say that "$x$ *can be reached from* $m$ *by* $\mathcal{O}$" provided that there exist $0_1, 0_2, \ldots, 0_n \in \mathcal{O}$ such that $[m, x] \subset 0_1 \cup 0_2 \cup \cdots \cup 0_n$. Then, let $A = \{x \in [m, M] | x$ can be reached from $m$ by $\mathcal{O}\}$ and let $t = \sup A (A \neq \emptyset$ since $m \in A)$. Since $\mathcal{O}$ covers $[m, M]$, there exists $0_t \in \mathcal{O}$ such that $t \in 0_t$. Therefore there are points $t', t'' \in 0_t$ such that $t' < t < t''$. Since $t' \in A$, there exist open intervals $0_1, \ldots, 0_k \in \mathcal{O}$ such that $[m, t'] \subset 0_1 \cup \cdots \cup 0_k$. Then, $[m, t''] \subset 0_1 \cup \cdots \cup 0_k \cup 0_t$ which forces $M < t''$ and yields that $\{0_1, \ldots, 0_k, 0_t\}$ is a finite subcover of $\mathcal{O}$. This does the trick for $E^1$.

Finally, (d) implies (a) in $E^n$: Let $C$ be any closed and bounded subset of $E^n$. Then, there exists $r > 0$ such that $C \subset B(\bar{a}, r)$, for some $\bar{a} = (a_1, a_2, \ldots, a_n) \in C$.

Therefore, $C \subset \Pi_{i=1}^{n}[a_1 - r, a_1 + r]$, which implies that $C$ is compact, because of Theorem 8 and Lemma 6(a).

As immediate consequences of the preceding results, we get some of the classical and invaluable results of analysis:

**10. Theorem.** Let $C$ be a compact subset of $E^n$ and $f : C \to E^m$ a continuous function. Then $K = f(C)$ is compact and, for each $b \in \partial K$, there exists $b' \in C$ such that $f(b') = b$. (If $m = 1$, this says that every continuous real-valued function $f : C \to E^1$ attains a maximum and a minimum value.)

**Proof.** Since $K$ is closed in $E^m$, by Theorem 9(d), $b \in \partial K$ implies that $b \in K$; therefore, there exists $b' \in C$ such that $f(b') = b$.

**11. Definition.** Let $(X, d)$ and $(Y, \rho)$ be metric spaces. We say that

  (a) A function $f : X \to Y$ is *uniformly continuous* (with respect to the metrics $d$ and $\rho$, of course) provided that, for each $\varepsilon > 0$, there exists $\delta > 0$ ($\delta$ depends only on $\varepsilon$) such that $\rho(f(x), f(w)) < \varepsilon$ whenever $d(x, w) < \delta$. (See ex. 4.)

  (b) For any $\delta > 0$, the *$\delta$-modulus of continuity of $f$*, denoted by $w(f, \delta)$, is $w(f, \delta) = \sup\{\rho(f(x), f(w))|d(x, w) \leq \delta\}$. (Of course, it can happen that $w(f, \delta) = +\infty$.)

**12. Lemma.** A function $f : (X, d) \to (Y, \rho)$ is uniformly continuous *iff* $\lim_{\delta \to 0} w(f, \delta) = 0$.

**Proof.** Straightforward.

**13. Theorem.** Let $(X, d)$ be a compact metric space, $(Y, \rho)$ any metric space, and $f : X \to Y$ a continuous function. Then $f$ is uniformly continuous.

**Proof.** Let $\varepsilon > 0$. For each $x \in X$, pick a neighborhood $B(x, \delta_x)$ of $x$ such that

$$f(B(x, 2\delta_x)) \subset B\left(f(x), \frac{\varepsilon}{2}\right).$$

Since $\{B(x, \delta_x)|x \in X\}$ is an open cover of the compact space $X$, let us say that $\{B(x_1, \delta_{x_1}), \ldots, B(x_m, \delta_{x_m})\}$ also covers $X$. Let $\delta = \min\{\delta_{x_1}, \ldots, \delta_{x_m}\}$. It follows that, for each $x \in X$, $B(x, \delta) \subset$ some $B(x_i, 2\delta_{x_i})$: indeed, $x \in B(x_i, \delta_{x_i})$ implies $B(x, \delta) \subset B(x_i, 2\delta_{x_i})$. Therefore, for each $x \in X$,

$$f(B(x, \delta)) \subset B(f(x), \varepsilon),$$

which completes the proof.

## 3.2   Compactification (One-Point Only)

In Proposition 2.17, we proved that, for $n = 1, 2, \ldots$,

$$B^n / S^{n-1} \simeq S^n.$$

Let us interpret this result in the light of the knowledge we have acquired since then: From the proof of Theorem 2.17, it is immediate that

$$S^n - \{\text{north pole}\} \simeq (B^n)^0 = B(\bar{0}, 1).$$

Since $S^n$ is compact, it is accurate to say that $S^n$ is the *smallest compact space which contains the space* $(B^n)^0$ *as a subspace.*

A close analysis and general description of this situation will follow.

### 14. Definition.

(a) A subspace $X$ of a space $Y$ is said to be *dense* in $Y$ provided that $X^- = Y$.

(b) The space $Y$ is a *compactification* of the space $X$ provided that $Y$ is compact and $X$ is homeomorphic to a dense subspace of $Y$.

For any space $(X, \tau)$ pick a point not in $X$, generally denoted by $\infty$. Let $\hat{X} = X \cup \{\infty\}$ and

$$\hat{\tau} = \tau \cup \{U \cup (\hat{X} - K) | U \in \tau, K \subset X, K \text{ is closed and compact}\}.$$

### 15. Lemma. For any space $(X, \tau)$, $(\hat{X}, \hat{\tau})$ is a topological space and $(X, \tau)$ is a subspace of $(\hat{X}, \hat{\tau})$. $(X, \tau)$ is dense in $(\hat{X}, \hat{\tau})$ iff $(X, \tau)$ is not compact.

**Proof.** To prove that $\hat{\tau}$ is a topology, it is clearly sufficient to verify that

$$\bigcap_{i=1}^{n} (U_i \cup (\hat{X} - K_i)) = \hat{X} - \bigcup_{i=1}^{n} (K_i - U_i),$$

$$\bigcup_{\alpha \in \Lambda} (U_\alpha \cup (\hat{X} - K_a)) = \bigcup_{\alpha \in \Lambda} U_\alpha \cup \left( \hat{X} - \bigcap_{\alpha \in \Lambda} K_\alpha \right)$$

(the first equality is very easily proved by contradiction; otherwise it can be difficult); the second equality is obvious; note that $K - U$ is compact whenever $K$ is compact and $U$ is open, because of Lemma 6. (Clearly, the finite union of compact spaces is compact and any intersection of compact spaces is also compact.)

Clearly $\tau = \hat{\tau}|X$ (note that $\hat{\tau}|X \subset \tau$, because in each $U \cup \hat{X} - K$, $K$ is closed).

### 16. Theorem. For any non-compact space $(X, \tau)$, $(\hat{X}, \hat{\tau})$ is a compactification of $(X, \tau)$. (It is called the *one-point compactification of* $X$, inasmuch that it is immediate that any two one-point compactifications of the same space are homeomorphic.)

**Proof.** By Lemma 15, we only need to show that $(\hat{X}, \hat{\tau})$ is compact: Let $\mathcal{U} \subset \hat{\tau}$ be an open cover of $\hat{X}$. Pick some $V \in \mathcal{U}$ with $\infty \in V$. Clearly

$$V = U \cup (\hat{X} - K),$$

since no $U \in \tau$ contains $\infty$. Since $\{U \cap X | U \in \mathcal{U}\} \subset \tau$ and $K$ is a compact subspace of $X$, there exists a finite subcollection $\{U_1, \ldots, U_n\}$ of $\mathcal{U}$ such that

$$K \subset \bigcup_{i=1}^{n} U_i$$

and thus $\{U_1, \ldots, U_n\} \cup \{V\}$ is a finite subcollection of $\mathcal{U}$ which covers $\hat{X}$. This completes the proof.

**17. Corollary.** For $n \in \mathbf{N}$, $S^n$ is the one-point compactification of $E^n$.

**Proof.** It is easy to see that in $E^n$,

$$B(\bar{0}, 1) \simeq E^n$$

and we already know that $S^n$ is the one-point compactification of $B(\bar{0}, 1) \subset E^n$. (Of course, $S^n \subset E^{n+1}$.)

The reader should not conclude from this corollary that the one-point compactification of a space, even a Hausdorff one, is Hausdorff. Such a conclusion is incorrect. The correct conclusion requires that we *localize* the notion of compactness.

**18. Definition.** A space $(X, \tau)$ is said to be *locally compact* provided that each $x \in X$ has a compact neighborhood.

**19. Lemma.** Let $(X, \tau)$ be Hausdorff. Then $(X, \tau)$ is *locally compact* if and only if, for each $x \in X$ and neighborhood $U$ of $x$, there exists a neighborhood $V$ of $x$ such that $\bar{V} \subset U$ and $\bar{V}$ is compact.

**Proof.** Immediate from Lemma 6.

**20. Theorem.** The space $(X, \tau)$ is locally compact Hausdorff if and only if $(\hat{X}, \hat{\tau})$ is compact Hausdorff.

**Proof.** Straightforward.

## 3.3 Complete Metric Spaces

In elementary calculus one becomes very aware of the usefulness of the result: *Every Cauchy sequence of real numbers converges to some real number.* This impressive result on Cauchy sequences is but a drop of water in a sea. Let us swim a little.

**21. Definition.** A sequence $\{x_n\}$ in a metric space $(X, d)$ is called a *Cauchy sequence* provided that, for each $\varepsilon > 0$, there exists an integer $N(\varepsilon)$ such that $m$,

$n > N(\varepsilon)$ implies $d(x_n, x_m) < \varepsilon$. A metric space $(X, d)$ is *complete* provided that each Cauchy sequence in $X$ converges in $X$.

**22.  Lemma.** Let $(X_i, d_i)$ be complete metric spaces for $i = 1, 2, \ldots, n$. Then $\Pi_{i=1}^n X_i$, with the product-metric, is a complete metric space.

**Proof.** Let $\{(x_1^k, x_2^k, \ldots, x_n^k)\}_k$ be a Cauchy sequence in $\Pi_{i=1}^n X_i$. Then it is easy to see that $\{x_i^k\}_k$ is a Cauchy sequence in $X_i$ for $i = 1, 2, \ldots, n$. Since each $(X_i, d_i)$ is complete, there exists $z_i \in X_i$, $i = 1, 2, \ldots, n$, such that $\lim_k x_i^k = z_i$, from which it easily follows that $\lim_k (x_1^k, x_2^k, \ldots, x_n^k) = (z_i, \ldots, z_n)$. This shows that $\Pi_{i=1}^n X_i$, with the product metric, is complete.

**23.  Lemma.** The real line $E^1$, with the Euclidean metric, is a complete metric space.

**Proof.** This follows easily from the local compactness of $E^1$ (see ex. 20). Intrinsic properties of the Euclidean metric are crucial (see ex. 21).

**24.  Corollary.** For each $n$, $E^n$ with the Euclidean metric, is a complete metric space.

**25.  Lemma.** A subspace $S$ of a complete metric space $(X, d)$ is complete if and only if $S$ is closed in $X$.

**Proof.** The "if" part is obvious. Let us look at the "only if" part: Let $p \in S^-$. Then there exists a sequence $\{x_n\} \subset S$ such that $d(x_n, p) \leq \frac{1}{n}$, for $n \in \mathbf{N}$. Then $\{x_n\}$ is a Cauchy sequence in $S$ which converges in $S$. Since $\lim_n X_n = p$, from Lemma 4, we get that $p \in S$. This shows that $S$ is closed.

Undoubtedly, the most applicable results of complete metric spaces are the *Baire Category Theorem* (cf. ex. 19) and *Banach's Contraction Theorem*, which we will now describe together with some elementary applications. Later on, there will be more to come (cf. Theorems 4.6 and 4.8).

**26. Definition.**

(a) Let $f : X \to X$ be any function. Then $p \in X$ is called a *fixed point of $f$* provided that $f(p) = p$.

(b) Let $(X, d)$ be a metric space and $0 \leq \alpha < 1$. If $f : X \to X$ satisfies the inequality

$$d(f(x), f(y)) \leq \alpha d(x, y)$$

for all $x$, $y \in X$, then $f$ is called an *$\alpha$-contraction*. If $d(f(x), f(y)) > \beta d(x, y)$, for all $x$, $y \in X$, and if $\beta > 1$, then the function $f$ is called a

$\beta$-expansion. If $d(f(x), f(y)) = d(x, y)$, for all $x$, $y \in X$, then $f$ is called a *d-isometry*, or just an *isometry* (when no confusion appears possible).

(c) For any function $f : X \to X$, inductively, we let $f^1 = f$, $f^2 = f \circ f^1, \ldots, f^n = f \circ f^{n-1}, \ldots$. The function $f^n : X \to X$ is called the *$n^{th}$-iterate of f.*

**27. Theorem.** (*Banach's contraction principle*). Let $(X, d)$ be a complete metric space and $f : X \to X$ be an $\alpha$-contraction, $0 \le \alpha < 1$. Then there exists a *unique* $q \in X$ such that $f(q) = q$. Furthermore, for each $x \in X$,

$$d(f^n(x), q) \le \frac{\alpha^n}{1 - \alpha} d(x, f(x)).$$

**Proof.** By induction, it is straightforward that

$$d(f^n(x), f^{n+1}(x)) \le \alpha^n d(x, f(x))$$

for any $x \in X$ and $n \in \mathbf{N}$. Therefore, for each $n < m$, a generalization of the triangle inequality yields

$$
\begin{aligned}
d(f^n(x), f^m(x)) &\le d(f^n(x), f^{n+1}(x)) + \cdots + d(f^{m-1}(x), f^m(x)) \\
&\le \alpha^n d(x, f(x)) + \cdots + \alpha^{m-1} d(x, f(x)) \\
&= \alpha^n (1 + \cdots + \alpha^{m-n-1}) d(x, f(x)) \\
&< \alpha^n (1 + \alpha + \cdots + \alpha^m + \cdots) d(x, f(x)) \\
&= \alpha^n \cdot \frac{1}{1 - \alpha} d(x, f(x)).
\end{aligned}
$$

Since $\lim_{n} \alpha^n = 0$ and $d(f^n(x), f^m(x)) < \frac{\alpha^n}{1-\alpha} d(x, f(x))$, it is easily seen that $\{f^n(x)\}$ is a Cauchy sequence. So let $q = \lim_{n} f^n(x)$.

First, note that $f(q) = q : q = \lim_{n} f^n(x)$ implies $f(q) = \lim_{n} f(f^n(x)) = \lim_{n} f^{n+1}(x) = q$, since any contraction is clearly a continuous function.

Finally, $q$ is *unique*: Suppose $p = f(p)$ and $p \ne q$. Then $d(p, q) = d(f(p), f(q)) \le \alpha d(p, q)$, implying that $d(p, q) < d(p, q)$, a contradiction. It is illuminating to observe that *regardless of which we X one starts with,*

$$\lim_{n} f^n(w) = q \text{ (the only fixed point of } f).$$

Finally, let us observe that, for each $x \in X$ and $n < j$,

$$
\begin{aligned}
d(f^n(x), q) &\le d(f^n(x), f^j(x)) + d(f^j(x), q) \\
&< \frac{\alpha^n}{1 - \alpha} d(x, f(x)) + d(f^j(x), q).
\end{aligned}
$$

Since $\lim_{j} f^j(x) = q$, we then get that

$$d(f^n(x), q) \le \frac{\alpha^n}{1 - \alpha} d(x, f(x)).$$

At this time we will treat the following two useful applications of the preceding result.

*Roots of $y = h(x)$.* Let $S = [a, b]$ and $h : S \to E^1$ be a differentiable function such that

(i) $h(a)h(b) < 0$,
(ii) there exists $m, M \in E^1$, such that $0 < m \leq |h'(x)| \leq M$, for each $x \in S$.

Then there exists a unique $a \leq q \leq b$ such that $h(q) = 0$: Let

$$f(x) = x - \lambda h(x) \text{ such that } 1 - \lambda M > 0.$$

Then $f : S \to S$ is a $(1 - \lambda m)$-contraction ($0 < f'(x)$ implies that $f$ is strictly increasing; therefore, $f(S) \subset S$, because $a < f(a)$ and $f(b) < b$; $|f'(x)| < 1 - \lambda m$ implies that $f$ is a $(1 - \lambda m)$-contraction by the Theorem of the Mean (if $a \leq x < y \leq b$ then $|\frac{f(x)-f(y)}{x-y}| = |f'(t)|$, for some $x \leq t \leq y$)). Therefore, for each $x \in S$

$$\lim_n f^n(x) = q \text{ such that } f(q) = q; \text{ hence } h(q) = 0.$$

Observe that this tells us exactly how to find the point $q$ to any desired degree of accuracy, since we know the rate of convergence of $\{f^n(x)\}$ to $q$.

*Systems of Linear Equations.* Let $A = (a_{ij})$ be a real $n \times n$-matrix and define $f : E^n \to E^n$ by $f(x) = Ax + b$, with $x$ and $b$ thought of as column vectors and $b$ fixed. When does $f$ have a unique fixed point? Certainly it suffices that $f$ be a contraction with respect to some complete metric on $E^n$. We will consider only three commonly used metrics.

(i) The Euclidean Metric: Note that

$$|f(\bar{x}) - f(\bar{y})|^2 = \Sigma_i(\Sigma_j a_{ij}(x_j - y_j))^2$$
$$\leq \Sigma_i(\Sigma_j a_{ij}^2)(\Sigma_j(x_j - y_j)^2)$$
$$= \Sigma_i(\Sigma_j a_{ij}^2)|\bar{x} - \bar{y}|^2$$

by Cauchy-Schwarz inequality (see Appendix A); therefore

$$|f(\bar{x}) - f(\bar{y})| \leq [(\Sigma_{i,j} a_{ij}^2)]^{1/2}|x - y|$$

and $f$ is a contraction with respect to the Euclidean metric whenever

$$\Sigma_{i,j} a_{ij}^2 < 1.$$

(ii) The metric $d_1(\bar{x}, \bar{y}) = \sup\{|x_1 - y_1|, \dots, |x_n - y_n|\}$: It is easy to see that $(E^n, d_1)$ is a complete metric space, and that $f$ is a contraction with respect to $d_1$ whenever

$$\Sigma_j|a_{ij}| < 1, \text{ for } i = 1, 2, \dots, n.$$

(iii) The metric $d_2(\bar{x}, \bar{y}) = \Sigma_i|x_i - y_i|$: It is easy to see that $(E^n, d_2)$ is a complete metric space, and that $f$ is a contraction with respect to $d_2$ whenever

$$\Sigma_i|a_{ij}| < 1, \text{ for } j = 1, 2, \dots, n.$$

We will conclude our introductory study of complete spaces with a beautiful and very useful characterization of compact metric spaces: We will call a metric space $(X, d)$ *totally bounded* provided that, for each $\varepsilon > 0$, there exists a finite set $F_\varepsilon \subset X$ such that $X = \bigcup\{B(u, \varepsilon) | u \in F_\varepsilon\}$.

**28. Theorem.** A metric space $(X, d)$ is compact iff $(X, d)$ is complete and totally bounded.

**Proof.** The "only if" part is straightforward. Because of Theorem 9, the "if" part becomes obvious once one proves that every infinite sequence $\{x_n\}$ of $X$ has a Cauchy subsequence: For $k = 1, 2, \ldots$, let $F_k$ be a $\frac{1}{k}$-net of $X$. Since $X = \bigcup\{B(u, 1) | u \in F_1\}$, there exists a subsequence $\{x_n^1\}$ of $\{x_n\}$ which is contained in some $B(u_1, 1)$, with $u_1 \in F_1$. Similarly, there exists a subsequence $\{x_n^2\}$ of $\{x_n^1\}$ which is contained in some $B(u_2, 1/2)$, with $u_2 \in F_2$. Inductively, we pick sequences $\{x_n^j\}$, for $j \in \mathbb{N}$, such that

   (i) $\{x_n^1\}$ is a subsequence of $\{x_n\}$
   (ii) $\{x_n^{j+1}\}$ is a subsequence of $\{x_n^j\}$ for $j = 1, 2, \ldots$
   (iii) $\{x_i^j\} \subset$ some $B(u_j, \frac{1}{j})$ with $u_j \in F_j$.

It follows that $\{x_n^n\}$ is a Cauchy subsequence of $\{x_n\}$, since

$$d(x_{n+p}^{n+p}, x_n^n) \leq d(x_{n+p}^{n+p}, u_n) + d(u_n, x_n^n) < \frac{1}{2^n} + \frac{1}{2^n} = \frac{1}{2^{n-1}}$$

for every $n$, $p$.

**29. Corollary.** Every compact metric space $X$ is separable (*i.e.*, there exists countable $D \subset X$ such that $D^- = X$).

**Proof.** Because of Theorem 28, let $F_n$ be a $\frac{1}{n}$-net for $n \in \mathbb{N}$. Let $D = \bigcup_{n=1}^{\infty} F_n$. It follows easily that $D$ is a countable dense (*i.e.*, $\bar{D} = X$) subset of $X$. This does it.

## 3.4 Connected and Arcwise Connected Spaces

It is in this context that the Intermediate-Value Theorem of Calculus finds its true meaning. For convenience, we call a subset $A$ of $(X, \tau)$ *clopen* whenever $A$ is both *an open and a closed subset* of $(X, \tau)$.

**30. Definition.** A space $(X, \tau)$ is called

   (i) *connected* provided that the only clopen subsets of $(X, \tau)$ are $X$ and $\emptyset$.
   (ii) *arcwise connected* provided that, for all $x$, $y \in X$, there exists a continuous function $\gamma : I \to X$ such that $\gamma(0) = x$ and $\gamma(1) = y$. The function $\gamma$ is called an *arc* or *path* starting at $x$ and ending at $y$. If $x = y$ then $\gamma$ is called a *loop with base point $x$*.

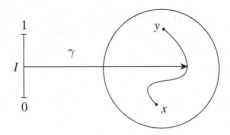

While these two notions appear to be the same for open subsets of Euclidean spaces—and they are indeed—they are, in general, quite different (see ex. 1).

**31.  Lemma.** Let $X$ be (arcwise) connected and $f : X \twoheadrightarrow Y$ be a continuous function. Then $Y$ is (arcwise) connected.

**Proof.** Assume $Y$ is not connected. Then there exists a clopen $B \subset Y$ such that $\emptyset \neq B \neq Y$. Then $f^{-1}(B)$ is clopen in $X$, with $\emptyset \neq f^{-1}(B) \neq X$, a contradiction.

Now, take any two points $y_1, y_2 \in Y$ and pick $x_1, x_2 \in X$ such that $f(x_1) = y_1$, $f(x_2) = y_2$. Let $\gamma$ be any arc starting at $x_1$ and ending at $x_2$. Then $f \circ \gamma$ is an arc starting at $y_1$ and ending at $y_2$.

**32. Definition.** A cover, $\mathcal{U}$ of a space $(X, \tau)$ is called *chainable* provided that, for all $U, V \in \mathcal{U}$, there exists a finite set $\{U_1, \ldots, U_n\} \subset \mathcal{U}$ such that $U = U_1$, $V = U_n$ and $U_i \cap U_{i+1} \neq \emptyset$ for $i = 1, \ldots, n-1$ (*i.e.*, $\{U_1, \ldots, U_n\}$ is a *chain linking $U$ to $V$*).

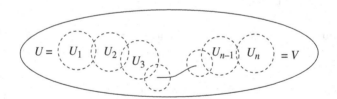

**33. Theorem.** Let $(X, \tau)$ be any space and $\mathcal{B}$ be any family of connected subspaces of $X$ such that $\mathcal{B}$ is a chainable cover of $Y = \bigcup \mathcal{B}$. Then $Y$ is connected.

**Proof.** Let $O$ be a nonempty clopen subset of $Y$. We will show that $O = Y$. First observe that if $B \in \mathcal{B}$ and $B \cap O \neq \emptyset$ then $B \subset O$ (otherwise, $B \cap O \neq B$ would be a nonempty clopen subset of $B$, a contradiction). Next observe that if $\{B_1, \ldots, B_n\}$ is a chain in $\mathcal{B}$ and $B_1 \subset O$ then $B_2 \subset O, \ldots, B_n \subset O$ ($B_2 \subset O$ because $B_2 \cap O \supset B_1 \cap B_2 \neq \emptyset$); similarly and inductively, $B_3 \subset O, \ldots, B_n \subset O$). Finally, since $\mathcal{B}$ is chainable and $O$ is nonempty, it follows that $O \supset \bigcup \mathcal{B} = Y$.

**34. Theorem.** A space $(X, \tau)$ is connected iff every open cover of $X$ is chainable.

**Proof.** The "if" part: Suppose $X$ is not connected. Pick a clopen subset $U$ of $X$ such that $\emptyset \neq U \neq X$. Then $\{U, X - U\}$ is an open cover of $X$ which is not chainable, a contradiction.

The "only if" part: Let $\mathcal{U}$ be any open cover of $X$ and define a relation $R$ on $\mathcal{U}$, by letting $URV$ provided that there exists a chain in $\mathcal{U}$ linking $U$ to $V$. It is straightforward that $R$ is an equivalence relation on $\mathcal{U}$. For each $U \in \mathcal{U}$ let

$$U_* = \bigcup \{V \in U | URV\}.$$

It is clear that $U_* \cap V_* \neq \emptyset$ iff $U_* = V_*$, for all $U, V \in \mathcal{U}$. Therefore, for each $U \in \mathcal{U}$, $U_* = X$ (otherwise $X - U$ is a union of sets $V_*$ with $V \in \mathcal{U}$, implying that $X - U$ is open or, equivalently, that $U_*$ is clopen and $\emptyset \neq U_* \neq X$, which contradicts the connectedness of $X$). This shows that $\mathcal{U}$ is chainable.

This most useful characterization of connectedness is generally mentioned as an after-thought. Among its many applications, the following is quite interesting.

**35. Lemma.** If an open subset $W$ of $E^n$ is connected then $W$ is arcwise connected.

**Proof.** Let $\mathcal{U}$ be a cover of $W$ consisting of open balls contained in $W$. Pick any points $x$, $y \in W$. By Theorem 34, there exists a chain $\{U_1 \ldots, U_m\}$ with $x \in U_1$ and $y \in U_m$. Geometrically, the remainder of the proof is trivial:

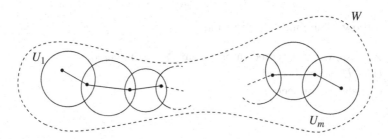

Indeed, not only can we construct an arc starting at $x$ and ending at $y$, but we can even construct one consisting of $m$ linear segments (such arcs are generally called *polygonal arcs*). Since the lengthy technical details are not commensurate with the triviality of this situation, we will omit them.

**36. Lemma.** Let $(X_1, \tau_1), \ldots, (X_n, \tau_n)$ be connected spaces. Then $(\Pi_{i=1}^N X_i, \Pi\tau_i)$ is connected.

**Proof.** Note that $(\Pi_{i=1}^{n-1} X_i) \times X_n \equiv \Pi_{i=1}^n X_i$, for each $n$. Therefore, by induction, it suffices to show that, for any two connected spaces $X$ and $Y$, $X \times Y$ is connected: Simply observe that, for some fixed $u \in Y$,

$$\{X \times \{u\}\} \cup \{\{x\} \times Y | x \in X\}$$

is a chainable (note that $\{x\} \times Y \cap X \times \{u\} = \{(x, u)\}$) cover of $X \times Y$ by connected subspaces (cf. ex. 2.18(b)). Therefore, by Theorem 33, $X \times Y$ is connected.

The preceding result can be easily generalized to state that any cartesian product of connected spaces is connected. The scheme of proof, which we present in ex. 14 is quite elucidative of a fruitful approach to the study of infinite products.

**37. Lemma.** Let $\{(X_\alpha \tau_\alpha)\}_{\alpha \in \Lambda}$ be any family of arcwise connected spaces. Then $(\Pi_{\alpha \in \Lambda} X_\alpha, \Pi\tau_\alpha)$ is arcwise connected.

**Proof.** Pick any $f, g \in \Pi_{\alpha \in \Lambda} X_\alpha$. For each $\alpha \in \Lambda$, there exists an arc $\psi_\alpha : I \to X_\alpha$ such that $\psi_\alpha(0) = f(\alpha)$ and $\psi_\alpha(1) = g(\alpha)$. Define $\psi : I \to \Pi_{\alpha \in \Lambda} X_\alpha$ by $[\psi(s)](a) = \psi_\alpha(s)$, for each $\alpha \in \Lambda$. Then $\psi(0) = f$, $\psi(1) = g$ and $\psi$ is continuous, because of Theorem 2.3 and Proposition 2.8. Therefore, $\psi$ is an arc starting at $f$ and ending at $g$.

**38. Theorem.** For every $a, b \in E^1$, with $a < b$, $[a, b]$ is connected.

**Proof.** Suppose not. Let $U$ be a clopen subset of $[a, b]$ such that $\emptyset \neq U \neq [a, b]$. Without loss of generality, let us assume that $b \notin U$ (if $b \in U$, then $b \notin U' = [a, b] - U$ and $U'$ is also clopen, $\emptyset \neq U' \neq [a, b]$). Now, let $s = \sup U$. Since $U$ is open in $[a, b[$, $s \notin U$. Since $U$ is closed in $[a, b]$, $s \in U$. We have a contradiction, which completes the proof.

**39. Corollary.** The following are valid:

   (a) $X$ is arcwise connected implies that $X$ is connected,
   (b) each $E^n$ is connected,
   (c) each $S^n$ is connected.

**Proof.**

   (a) Suppose $X$ is not connected. Pick clopen $U \subset X$ such that $\emptyset \neq U \neq X$. Pick $x \in U$ and $y \in X - U$. Suppose that there exists an arc $\gamma : I \to X$ such that $\gamma(0) = x$ and $\gamma(1) = y$. Then $\gamma^{-1}(U \cap \gamma(I)) = \gamma^{-1}(U)$ is a clopen subset of $I$, with $\emptyset \neq \gamma^{-1}(U) \neq I$, a contradiction.
   (b) Because of Lemma 36, it suffices to prove that $E^1$ is connected. Since $E^1 = \bigcup_{n=1}^{\infty} [-n, n]$, it follows, from Theorems 33 and 38, that $E^1$ is connected.
   (c) First define $r : E^{n+1} - \{\bar{0}\} \to S^n$ by $r(x) = \frac{x}{|x|}$. It is clear that $r$ is a well-defined, onto and continuous function. Furthermore, it is easy to see that $E^{n+1} - \{\bar{0}\}$ is arcwise connected (to join the points $x$ and $y = -x$ in $E^{n+1} - \{\bar{0}\}$ use a semi-circle with center at $\bar{0}$, radius $|x|$, starting at $x$ and ending at $y$; that is, the secret is to go around $\bar{0}$), and therefore $E^{n+1}$ is connected. By Lemma 31, $S^n$ is connected for $n \in \mathbf{N}$.

**40. Theorem** (*Generalized* Intermediate-Value). Let $C$ be a compact connected subset of $E^n$ and $f : C \to E^1$ a continuous function. Then $f(C) = [a, d]$, with $a = \inf\{f(x)|x \in C\}$ and $d = \sup\{f(x)|x \in C\}$.

**Proof.** From Theorem 10, we get that $a, d \in f(C)$. Suppose that there exists $a < s < d$ such that $s \notin f(C)$. Then, letting

$$U = ]-\infty, s[ \bigcap f(C),$$

we get that $U$ is a clopen subset of $f(C)$, with $\emptyset \neq U \neq f(C)$, which contradicts Lemma 31.

## Chapter 3.   Exercises

1. Let $S$ be the subspace of $E^2$ which is the union of $A = \{(0, y)| -1 \leq y \leq 1\}$ and $B = \{(x, y)|y = \sin\frac{1}{x}, 0 < x \leq \pi\}$.

   Show that $S$ is connected but not arcwise connected. Indeed show that any arc which starts in $A$ stays in $A$. (Hint: Assume, for example, that there exists an arc $\gamma$ starting at $(0, 0)$ and ending at $(p, \sin\frac{1}{p})$. Let $t = \inf\{s \in I|\gamma(s) \in B\}$. Show that $\gamma(t) \notin B$, because of continuity of $\gamma$. Similarly, show $\gamma(t) \notin A$, because $\gamma(t) \in A$ implies that there exists $\delta > 0$ such that $\gamma(]t, t + \delta]) \subset B(\gamma(t), 1) - A$, and therefore that $\gamma(]t, t+\delta[)$ is not arcwise connected. We have a contradiction.)

2. Let $T$ be the subspace of $E^2$ which is the union of $\{(0, 0)\}$ with $\{(x, y)|y = x \sin\frac{1}{x}, 0 < x \leq \pi\}$.

   Show that $T$ is connected and arcwise connected.

3. Let $A$ be any connected subspace of $(X, \tau)$. Then $A^-$ is a connected subspace of $(X, \tau)$. (Hint: See ex. 1.5.)

4. Define $f :]0, 1[\to E^1$ by $f(x) = \frac{1}{x}$. Show that $f$ is not uniformly continuous. Indeed, show that $\lim_{\delta \to 0} w(f, \delta) = +\infty$.

5. A metric space $(X, d)$ is compact iff every continuous real-valued function on $X$ is bounded. (Hint: Assume $X$ is not compact. Show that there exists a sequence $\{x_n\}$ in $X$ and $\varepsilon_n > 0$ such that $B(x_n, \varepsilon_n) \cap \overline{\bigcup_{m \neq n} B(x_m, \varepsilon_m)} = \emptyset$, for each $n \in \mathbf{N}$. Next, define $f : X \to E^1$ by $f(x_n) = n$, $f(X - \bigcup_n B(x_n, \varepsilon_n)) = 0$ and $f(x) = (1 - \frac{d(x, x_n)}{\varepsilon_n})n$, for each $x \in B(x_n, \varepsilon_n)$. Show that $f$ is an unbounded continuous function.)

6. Let $X$ be a compact space and $z \in X$. Show that the connected component of $X$ containing $z$ is the intersection of all open and closed subsets of $X$ which contain $z$.

7. Show that any compact metric space is complete.

8. Show that $I$ is not homeomorphic to $S^1$.

9. Show that $S^n(n > 1)$ and $S^1$ are not homeomorphic.

10. Letting $J = ]0, 1[$, show that

(a) $J$ is homeomorphic to a subspace of $I$ and $I$ is homeomorphic to a subspace of $J$.

(b) $J$ and $I$ are *not* homeomorphic. (Compare this with the Schroeder-Bernstein Theorem on cardinality.)

11. Let $(X, d)$ be compact metric and $f : X \to X$ a function such that $d(f(x), f(y)) = d(x, y)$, for all $x, y \in X$. Show that $f$ is onto. (Hint: By contradiction)

12. A family $\mathcal{A}$ of subsets of a set $X$ is said to have the *finite intersection property* (*i.e.*, *fip*) provided that, for every finite $\mathcal{F} \subset \mathcal{A}$, $\bigcap \mathcal{F} \neq \emptyset$. For any space $(X, \tau)$ prove that the following are equivalent:

   (i) $X$ is compact,
   (ii) For every family $\mathcal{A}$ of closed subsets of $X$ with fip, $\bigcap \mathcal{A} \neq \emptyset$,
   (iii) For every family $\mathcal{A}$ of closed subsets of $X$ with $\bigcap \mathcal{A} = \emptyset$, there exits a finite $\mathcal{F} \subset \mathcal{A}$ such that $\bigcap \mathcal{F} = \emptyset$.

13. *Lebesgue Number.* Let $\mathcal{U}$ be an open cover of a metric space $(X, \rho)$. Any $\delta > 0$ such that, for each $x \in X$,

$$B(x, \delta) \subset \text{some } U \in \mathcal{U},$$

is called a *Lebesgue number* for $\mathcal{U}$.

   (i) Let $\mathcal{U} = \{]n - \frac{1}{n}, n + \frac{1}{n}[ | n = 1, 2 \ldots\} \cup \{E^1 - \mathbf{N}\}$. Show that $\mathcal{U}$ is an open cover of $E^1$ with no Lebesgue number.
   (ii) Show that every open cover $\mathcal{U}$ of a compact metric space $(X, \rho)$ does have Lebesgue numbers. (Hint: Let $\mathcal{B} = \{B(x, \varepsilon(x)) | B(x, 2\varepsilon(x)) \subset \text{some } U \in \mathcal{U}\}$. Let $\{B(x_1, \varepsilon(x_1)), \ldots, B(x_m, \varepsilon(x_m))\}$ be a finite subcover of $\mathcal{B}$ and let $\delta = \min\{\varepsilon(x_1), \ldots, \varepsilon(x_m)\}$. Show that $\delta$ is a Lebesgue number for $\mathcal{B}$ and, therefore, for $\mathcal{U}$.

14. Let $\{(X_\alpha, \tau_\alpha)\}_{\alpha \in \Lambda}$ be a family of connected spaces. Let $X = \Pi_{\alpha \in \Lambda} X_\alpha$ and $\tau = \Pi \tau_\alpha$. Pick any $\ell \in X$ and let (see ex. 2.18)

$$P(\ell) = \bigcup \{S(\ell, \Gamma) | \Gamma \subset \Lambda, \Gamma \text{ is finite}\}.$$

   (i) Show that $P(\ell)$ is dense in $X$ (see ex. 2.18).
   (ii) Show that each $S(\ell, \Gamma)$, such that $\Gamma \subset \Lambda$ and $\Gamma$ is finite, is connected (see ex. 2.18 and Lemma 36).
   (iii) Show that $\{S(\ell, \Gamma) | \Gamma \subset \Lambda, \Gamma \text{ is finite}\}$ is a chainable cover of $P(\ell)$. (Hint: $\ell \in$ each $S(\ell, \Gamma)$).
   (iv) Show that $P(\ell)$ is connected (cf. Theorem 33).
   (v) Show that $X$ is connected (see ex. 3).

15. Let $X = I \times I$ ordered by the *lexicographic order* (*i.e.*, $(a, b) < (c, d)$ iff $a < c$ or $a = c$ and $b < d$). Let $X$ have the order topology $\tau_0$ (see ex. 1.10) and show that

   (a) $(X, \tau_0)$ is compact Hausdorff.

(b) $(X, \tau_0)$ is connected.

(c) $(X, \tau_0)$ is not arcwise connected.

(d) If $Y = I \times \{1/2\}$ then $\tau_0 | Y$ is the discrete topology (see ex. 1.2).

(e) If $Z = I \times \{I\}$, then $\tau_p | Z$ is the half-open interval topology on $Z$ (see ex. 1.3).

(f) If $W = \{t\} \times I$, $t \in I$, then $\tau_0 | W$ is the Euclidean topology on $W$.

16. *Generalization of Theorem* 27. Let $(X, d)$ be a complete metric space and $f : X \to X$ a function such that $f^m$ is an $\alpha$-contraction. Show that $f$ has a unique fixed point. (Hint: Consider the sequence $\{f^{km}(x)\}_k$, for any $x \in X$. Note that $d(f^{km}f(x), f^{km}(x)) \le \alpha^k d(f(x), x)$. Let $q = \lim_k f^{km}(x)$. Show $f(q) = q$ and that $q$ is unique.)

17. *Localized Contraction Principle.* Let $(X, d)$ be a complete metric space and $f : X \to X$ be a function such that $f$ is an $\alpha$-contraction on $\overline{B(x_0, r_0)}$, where $\frac{1}{1-\alpha} d(x_0, f(x_0)) = r_0$. Then $f$ has a unique fixed point $p$ in $\overline{B(x_0, r_0)}$; furthermore, letting $x_1 = f(x_0), \ldots, x_{n+1} = f(x_n), \ldots$, we get that $d(x_m, p) \le \alpha^m r_0$. (Hint: Note that $d(x_1, x_0) = (1 - \alpha)r_0 < r_0$. Assume that $x_0, x_1, \ldots, x_n \in \overline{B(x_0, r_0)}$ such that $d(x_n, x_0) \le (1 - \alpha^n)r_0 < r_0$. Then show that $d(x_{n+1}, x_0) \le d(x_{n+1}, x_n) + d(x_n, x_0) \le \alpha^n d(x_1, x_0) + (1 - \alpha^n)r_0 \le \alpha^n(1-\alpha)r_0 + (1-\alpha^n)r_0 = (1 - \alpha^{n+1})r_0 < r_0$. By induction, this shows that the sequence $\{x_n\}$ is contained in the complete metric space $\overline{B(x_0, r_0)} \ldots$).

A topological space, $(X, \tau)$, is said to be

(i) *sequentially compact* if every sequence $\{x_n\}$ in $X$ has a convergent subsequence,

(ii) *countably compact* if every countable open cover $\mathcal{V}$ of $X$ has a finite subcover.

18. Prove the following

(a) Every sequentially compact space $(X, \tau)$ is countably compact.

(b) A metric space is compact iff it is sequentially compact iff it is countably compact.

19. *Baire Category Theorem.* Let $(X, d)$ be a complete metric space and $U_1, U_2, \ldots$ be a sequence of open dense subsets of $X$. Then $D = \bigcap_{n=1}^{\infty} U_n$ is a dense subset of $X$. (Hint: Pick $x \in X$ and a neighborhood $U$ of $x$. We need to show that $U \cap D \ne \emptyset$. We construct a Cauchy sequence $\{x_n\}$ in $X$ such that $z = \lim_n x_n$ and $z \in U \cap D$: Pick $B(x, \varepsilon_0) \subset \overline{B(x, \varepsilon_0)} \subset U$, such that $\varepsilon_0 \le 1$, and choose $x_1 \in B(x, \varepsilon_0) \cap U_1$. Pick $B(x_1, \varepsilon_1) \subset \overline{B(x_1, \varepsilon_1)} \subset B(x, \varepsilon_0) \cap U_1$, such that $\varepsilon_1 \le 1/2$ and choose $x_2 \in B(x_1, \varepsilon_1) \cap U_2$. Pick $B(x_2, \varepsilon_2) \subset \overline{B(x_2, \varepsilon_2)} \subset B(x_1, \varepsilon_1) \cap U_2$ such that $\varepsilon_2 < \frac{1}{2^2}$ and choose $x_3 \in B(x_2, \varepsilon_2) \cap U_3$. In this fashion, inductively choose $x_4, x_5, \ldots$. Check that the sequence $\{x_n\}$ is Cauchy and its limit is in $U \cap D$.

20. $(E^1,$ Euclidean metric$)$ is a complete metric space. (Hint: Let $\{x_n\}$ be a Cauchy sequence in $E^1$. Note that $\{x_n | n \in \mathbf{N}\} \subset$ some $[a, b]$ since there exists $t \in \mathbf{N}$

such that $m, n \geq t$ implies $|x_n - x_m| < 1$; hence, let $a = \min\{x_1, \ldots, x_t\} - 1$, $b = \max\{x_1, \ldots, x_t\} + 1$. Therefore $\{x_n\}$ has a convergent subsequence $\{x_{n_k}\}$; say $\lim\limits_{k} x_{n_k} = c$. Because $\{x_n\}$ is Cauchy, show that $\lim\limits_{n} x_n = c$.)

21. Prove that

   (a) the function $\tan^{-1} : E^1 \twoheadrightarrow \left]-\frac{\pi}{2}, \frac{\pi}{2}\right[$ is a homeomorphism;

   (b) if $d(x, y) = |\tan^{-1} x - \tan^{-1} y|$. then $d$ is a metric for $E^1$ such that $(E^1, \Gamma_1) \simeq (E^1, d)$ (recall $\Gamma_1$ is the Euclidean metric);

   (c) $\{n\}$ is $d$-Cauchy but does not converge; therefore, $(E^1, d)$ is not complete.

# Chapter 4

# Function Spaces

Many a colorful comment can be uttered about the importance of function spaces. Let us simply say that, after prospecting for so long, we will finally be rewarded with some of the most fabulous results of mathematics. And let there be no misunderstandings—much more lies elsewhere.

## 4.1 Function Space Topologies

For any spaces $(X, \tau)$ and $(Y, \mu)$, we let $C(X, Y) \equiv Y^X = \{f : X \to Y | f$ is continuous$\}$. Generally, the set $Y^X$ is called a *function space*. It is clear that $Y^X \subset \Pi_{x \in X} Y_X$, with each $Y_X = Y$; therefore, it can always be given the subspace topology with respect to any topology on $\Pi_{x \in X} Y_X$. But what we really want is to topologize $Y^X$ in such a way that we can generalize the various notions of "nearness" of functions $f, g : E^1 \to E^1$ which are generally studied in calculus. In one form or another, given $\varepsilon > 0$, it is understood that

    (i) $f$ and $g$ are '*uniformly $\varepsilon$-near*' provided that $|f(x) - g(x)| < \varepsilon$, for every $x \in E^1$

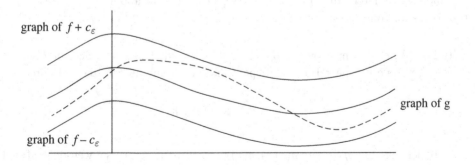

graph of $f + c_\varepsilon$

graph of g

graph of $f - c_\varepsilon$

    (ii) $f$ and $g$ are *$\varepsilon$-near on the interval* $[a, b]$ provided that $|f(x) - g(x)| < \varepsilon$, for $a \leq x \leq b$.

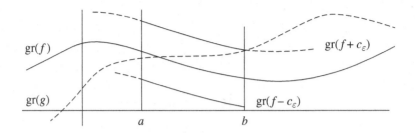

(iii) $f$ and $g$ are *$\varepsilon$-near at the point $p$* provided that $|f(p) - g(p)| < \varepsilon$.

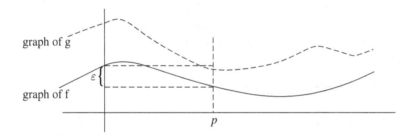

Various results are then proved, including the mainstay of Fourier Analysis: A *uniformly Cauchy* sequence $\{f_n\}$ of continuous functions $f_n : E^1 \to E^1$ (*i.e.* for every $\varepsilon > 0$, there exists $n(\varepsilon) \in \mathbf{N}$ such that $n, m > n(\varepsilon)$ implies $|f_n(x) - f_m(x)| < \varepsilon$, for all $x \in E^1$ (*i.e.* $f_n$ and $f_m$ are uniformly $\varepsilon$-near) *uniformly* converges to a continuous function $f : E^1 \to E^1$ (*i.e.* for every $\varepsilon > 0$ there exists $m(\varepsilon) \in \mathbf{N}$ such that $n > m(\varepsilon)$ implies $f_n$ is uniformly $\varepsilon$-near $f$). Later (cf. Lemma 2) we shall prove this result in a general setting.

There should remain no doubt that we need to talk about some metrics and topologies for function spaces $Y^X$:

(i) Let $(X, \tau)$ be any space and $(Y, d)$ be a boanded metric space, or else, let $(X, \tau)$ be a compact Hausdorff space and $(Y, d)$ any metric space. Define

$$d_s : Y^X \times Y^X \to E^1,$$

by letting $d_s(f, g) = \sup\limits_{x \in X} d(f(x), g(x))$—clearly the "sup" is a well-defined real number, in either case. It is straightforward to check that $d_s$ is a metric for $Y^X$, which is generally called the *sup-metric*. The topology generated by $d_s$ is generally called the *uniform-convergence* topology (abrev. *uc*).

Such is the utility of the sup metric, that it is customary to let $\{f_n\}$ is uniformly Cauchy $\equiv \{f_n\}$ is a Cauchy sequence with respect to the sup metric, $\{f_n\}$ converges uniformly to $f \equiv \lim_n d_s(f_n, f) = 0$.

(ii) Let $(X, \tau)$ and $(Y, \mu)$ be any topological spaces. For each $C \subset X$ and $U \subset Y$ let

$$\langle C, U \rangle = \{f \in Y^X | f(C) \subset U\}.$$

Then $\mathcal{S}_{co} = \{\langle C, U \rangle | C \text{ is compact}, U \text{ is open}\}$ is a subbase for a topology on $Y^X$ which is called the *compact-open topology* (abrev. *co*). The collection $\mathcal{S}_{pc} = \{\langle \{x\}, U \rangle | x \in X, U \text{ open}\}$ is a subbase for a topology on $Y^X$ which is called the *pointwise-convergence* topology (abrev. *pc*).

It should be rather obvious that the *uc* topology is closely related to the notion of "uniformly $\varepsilon$-near", the *co* topology is closely related to the notion of "$\varepsilon$-near on an interval $[a, b]$", and the *pc* topology is closely related to the notion of "$\varepsilon$-near at a point $x$". The following facts are also not hard to prove.

**1. Lemma.** The following are valid:

(a) For any spaces $X$ and $Y$, $pc \subset co$ on $Y^X$.

(b) For any spaces $X$ and $Y$, $(Y^X, pc)$ is a subspace of $\Pi_{x \in X} Y_X$ with the product topology (where each $Y_X = Y$).

(c) For any space $X$ and bounded metric space $(Y, d)$, $pc \subset co \subset uc$.

(d) For any compact Hausdorff space $X$ and metric space $(Y, \rho)$, $co = uc$.

**Proof.** Part (a) is obvious, since $\mathcal{S}_{pc} \subset \mathcal{S}_{co}$.

(b) Note that $\langle \{x\}, U \rangle = \Pi_x^{-1}(U) \cap Y^X$. This does the trick.

(c) Clearly $pc \subset co$. To show that $co \subset uc$ it suffices to show that each $\langle C, U \rangle \in \mathcal{S}_{co}$ is an element of $uc$: So let $f \in \langle C, U \rangle$. Then $d(f(C), Y - U) = \varepsilon > 0$. Therefore $f \in B(f, \varepsilon) \subset \langle C, U \rangle$ (suppose that there exists $g \in B(f, \varepsilon)$ such that $g(C) \not\subset U$. Pick $c \in C$ such that $g(c) \in X - U$. Then $d_s(f, g) \geq \varepsilon$, a contradiction). This shows that each $\langle C, U \rangle$ is a union of elements of $uc$, and therefore that $co \subset uc$.

(d) Because of (c), we only need to show that $uc \subset co$, for which it suffices to show that any $B(f, \varepsilon) \in co$: Geometrically, if one thinks of $B(f, \varepsilon)$ as a highway with center line, the graph of $f$, and thinks of $\langle C, U \rangle \in \mathcal{S}_{co}$ as a rectangle with compact base and open height, then all one wants is to trap any trajectory $g$ on the highway $B(f, \varepsilon)$ with a finite chain of rectangles.

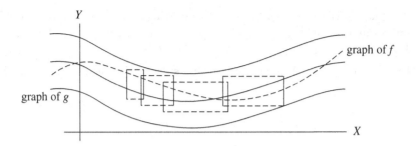

Analytically, let $g \in B(f, \varepsilon)$. Say $d_s(f, g) = \varepsilon - \delta$ with $\delta > 0$. Then, for each $x \in X$, $B(g(x), \delta) \subset B(f(x), \varepsilon)$ and, since $g$ is continuous and $X$ compact Hausdorff, there exists a compact neighborhood $C_x$ of $x$ such that $g(C_x) \subset B(g(x), \delta)$. Since $\{(C_x)^0 | x \in X\}$ is an open cover of $X$, let $\{(C_{x_1})^0, \ldots, (C_{x_n})^0\}$ be a finite subcover of $X$. Then, it is easily seen that

$$g \in \bigcap_{i=1}^{n} \langle C_x, B(g(x_i), \delta) \rangle \subset B(f, \varepsilon),$$

which shows that $B(f, \varepsilon)$ is a union of elements of a base for $co$, and therefore that $B(f, \varepsilon) \in co$, which completes the proof.

## 4.2    Completeness and Compactness

**2. Lemma.** Let $X$ be any space and $(Y, d)$ a bounded and complete metric space (or $X$ a compact space and $(Y, d)$ any complete metric space). Then $(Y^X, \rho)$ is complete whenever $\rho$ is

   (i) the sup-metric $d_s$,
  (ii) the *weighted-metric* $d_w$ defined by

$$d_w(f, g) = \sup_X e^{-\delta d(x, x_0)} d(f(x), g(x)),$$

     where $x_0$ is a fixed point of $X$ and $\delta > 0$,
 (iii) the *integral-metric* $d_i$ defined by

$$d_i(f, g) = \int_X |f(x) - g(x)| dx,$$

     whenever this makes sense (for example, whenever $X$ is a closed interval of $E^1$ and $Y = E^1$).

**Proof.** We will deal only with the metric $d_s$, since the other cases are quite similar (the reader should check that $d_w$ and $d_i$ are actually metrics—the fact that $Y^X$ consists of continuous functions is crucial in showing that $d_i$ is a metric). So let $\{f_n\}$ be a Cauchy sequence in $(Y^X, d_s)$. Then, for each $x \in X$, $\{f_n(x)\}$ is obviously a Cauchy sequence. This enables us to define a function $f : X \to Y$ by

$$f(x) = \lim_n f_n(x),$$

for each $x \in X$. We will show that $f$ is continuous and that $\lim_n f_n = f$. First, we show that

(i) for every $\varepsilon > 0$, there exists $n(\varepsilon) \in \mathbf{N}$ such that $n > n(\varepsilon)$ implies $d(f_n(x), f(x)) \leq \varepsilon/3$ for all $x \in X$ (note that it is incorrect to deduce from this that $d_s(f_n, f) \leq \varepsilon/3$ until we have shown that $f$ is continuous): Pick $n(\varepsilon)$ such that $n, m > n(\varepsilon)$ implies $d(f_n(x), f_m(x)) < \varepsilon/3$, for all $x \in X$ (i.e. $d_s(f_n, f_m) \leq \varepsilon/3$). Then $d(f_n(x), f(x)) \leq d(f_n(x), f_m(x)) + d(f_m(x), f(x))$ for $m > n(\varepsilon)$. Therefore, for every

$$x \in X, \lim_m d(f_n(x), f(x)) \leq \lim_m d(f_n(x), f_m(x)) + \lim_m d(f_m(x), f(x)),$$

which implies that

$$d(f_n(x), f(x)) \leq \varepsilon/3, \text{ for every } x \in X.$$

Next, we show that $f$ is continuous at each $q \in X$: Simply observe that

$$d(f(q), f(y)) \leq d(f(q), f_n(q)) + d(f_n(q), f_n(y)) + d(f_n(y), f(y))$$

for every $n \in \mathbf{N}$. Using (i) and the continuity of $f_n$, we immediately get that $f$ is continuous.

Finally we show that $\lim_n f_n = f$: Immediate from (i) and the continuity of $f$.

Even if $X = Y = I$, we still cannot claim that $(Y^X, d_s)$ is compact. (Let $f : I \to I$ be defined by $f(x) = x^2$. Then the sequence $\{f^n\}$ has no convergent subsequence: Simply, note that, for any subsequence $\{f^{n_k}\}$ of $\{f^n\}$ we have that

$$\lim_k f^{n_k}(x) = \lim_k x^{2n_k} = \begin{cases} 0, & \text{for } 0 \leq x < 1, \\ 1, & \text{for } x = 1, \end{cases}$$

which shows that $\{f^{n_k}\}$ does not converge in $(I^I, d_s)$.) Therefore, given the many uses of uniformly convergent sequences of continuous functions, some of which will appear shortly, it becomes imperative to determine which subsets of $(Y^X, d_s)$ are compact.

**3. Definition.** A family $\mathcal{F}$ of $(Y^X, d_s)$ is said to be *equicontinuous at the point* $q \in X$ provided that, for every $\varepsilon > 0$, there exists a neighborhood $N_q$ of $q$ in $X$ such that

$$f(N_q) \subset B(f(q), \varepsilon),$$

for each $f \in \mathcal{F}$ (roughly speaking, all $f \in \mathcal{F}$ are equally continuous). The family $\mathcal{F}$ is said to be *equicontinuous* (on $X$) if $\mathcal{F}$ is equicontinuous at each point $q \in X$.

**4. Lemma.** If $\mathcal{F}$ is an equicontinuous subset of $(Y^X, d_s)$ then so is $\bar{\mathcal{F}}$ equicontinuous.

**Proof.** Pick any $x \in X$ and $\varepsilon > 0$, and let $N_x$ be a neighborhood of $x$ such that $f(N_x) \subset B(f(x), \varepsilon/3)$, for each $f \in \mathcal{F}$. We will show that

$$h(N_x) \subset B(h(x), \varepsilon), \text{ for each } h \in \bar{\mathcal{F}}.$$

Note that, for any $h \in \bar{\mathcal{F}}$ and $g \in \mathcal{F}$,

$$d(h(x), h(y)) \leq d(h(x), g(x)) + d(g(x), g(y)) + d(g(y), h(y)).$$

Pick any function $g \in \mathcal{F}$ such that $d_s(h, g) < \varepsilon/3$. It follows that $h(N_x) \subset B(h(x), \varepsilon)$. Since this is true for any $h \in \bar{\mathcal{F}}$, the proof is complete.

**5. Theorem (Ascoli-Arzela).** Let $X$ be a compact metric space, $(Y, d)$ any metric space and $\mathcal{F}$ a subset of $(Y^X, d_s)$. Then $\mathcal{F}$ is compact iff the following two conditions are valid:

   (i) $\mathcal{F}$ is equicontinuous.
   (ii) For each $x \in X$, $Z_x = \{f(x) | f \in \mathcal{F}\}^-$ is a compact subspace of $Y$.

**Proof.** The "if" part: Because of Lemma 4 and the obvious fact that $Z_x = \{f(x) | f \in \bar{\mathcal{F}}\}^-$. We assume, without loss of generality, that $\mathcal{F}$ is closed and infinite. Let $\{f_i\}$ be a sequence in $\mathcal{F}$ and let $D = \{x_j\}$ be a dense subset of $X$, by Corollary 3.29.

We will first show that there exists a subsequence $\{g_i\}$ of $\{f_i\}$ such that $\lim_i g_i(x_j)$ exists for each $x_j$: Since each $Z_{x_k}$ is compact, by induction, it is easy to pick sequences $\{f_i^k\}$ for $k \in \mathbf{N}$ such that

   (i) $\{f_i^I\}$ is a subsequence of $\{f_i\}$,
   (ii) $\{f_i^{k+I}\}$ is a subsequence of $\{f_i^k\}$,
   (iii) $\{f_i^k(x_j)\}$ converges for each $x_j$ with $j \leq k$.

Now, letting $g_i = f_i^i$, for $i \in \mathbf{N}$, it is easy to see that $\{g_i(x_j)\}_{i=1}^\infty$ converges, for each $x_j$ (note that $f_i^i$ contains a subsequence of each $f_i^k$, and apply Proposition 3.2(a)).

Next we show that $\{g_i(x)\}_i$ converges for each $x \in X$: Pick sequence $\{z_n\} \subset D$ such that $\lim_n z_n = x$ and note that

$$d(g_i(x), g_j(x)) \leq d(g_i(x), g_i(z_n)) + d(g_i(z_n), g_j(z_n)) + d(g_j(z_n)g_j(x)),$$

with each of the summands tending to zero as $i$, $j$, $n$ become large. Therefore $\{g_i(x)\}$ is a Cauchy sequence in the compact space $Z_x$ which implies that $\{g_i(x)\}$ converges.

Now define $f : X \to Y$ by letting

$$f(x) = \lim_i g_i(x),$$

for each $x \in X$. First, we prove that $f$ is continuous: Pick $q \in X$ and $\varepsilon > 0$. By equicontinuity of $\mathcal{F}$, there exists a neighborhood $N_q$ of $q$ such that $g_i(N_q) \subset B(g_i(q), \varepsilon/2)$, for $i \in \mathbf{N}$. Since

$$d(f(q), f(z)) \leq d(f(q), g_i(q)) + d(g_i(q), g_i(z)) + d(g_i(z), f(z)),$$

with $\lim_i g_i(q) = f(q)$ and $\lim_i g_i(z) = f(z)$, one immediately gets that

$$f(N_q) \subset \overline{B(f(q), \varepsilon|2)} \subset B(f(q), \varepsilon).$$

Finally, we prove that $\lim_i d_s(f, g_i) = 0$ (i.e. $\lim_i g_i = f$ in $(Y^X, d_s)$): Suppose not. Then there exists $\varepsilon > 0$ and subsequence $\{h_i\}$ of $\{g_i\}$ such that $d_s(f, h_i) \geq \varepsilon$, for $i \in \mathbf{N}$. Therefore, for every $i \in \mathbf{N}$ there exists $x_i \in X$ such that $d(f(x_i), h_i(x_i)) \geq \varepsilon/2$. Let $x$ be a cluster point of the sequence $\{x_i\}$ (note that $\{x_i\}$ may be finite). By equicontinuity of $\mathcal{F}$, there exists a neighborhood $N_x$ of $x$ such that $h_i(N_x) \subset B(h_i(x), \varepsilon/8)$. Since

$$\varepsilon/2 \leq d(f(x_i), h_i(x_i)) \leq d(f(x_i), f(x)) + d(f(x), h_i(x)) + d(h_i(x), h_i(x_i)),$$

with $\lim_i f(x_i) = f(x)$ (because $f$ is continuous) and $\lim_i h_i(x) = f(x)$ (because $\{h_i\}$ is a subsequence of $\{g_i\}$), we immediately get that, for some sufficiently large $m \in \mathbf{N}$,

$$x_m \in N_x, d(f(x_m), f(x)) < \frac{\varepsilon}{8}, d(f(x), h_m(x)) < \frac{\varepsilon}{8}.$$

It follows that

$$\frac{\varepsilon}{2} \leq d(f(x_m), h_m(x_m)) \leq \frac{\varepsilon}{8} + \frac{\varepsilon}{8} + \frac{\varepsilon}{8} < \frac{\varepsilon}{2},$$

a contradiction. The proof of the "if" part is complete.

The proof of the "only if" part is left as an exercise (see ex. 10).

In many aspects, Theorem 5 can be easily generalized to metric spaces $X$ for which there exists a sequence $\{C_n\}$ of compact subspaces with $X = \bigcup_{n=1}^{\infty} C_n$ (this includes all Euclidean Spaces $E^n$). See ex. 18 for details.

We end this section with three applications of Banach's Contraction Theorem, Theorem 5 and the Baire Category Theorem.

**6. Theorem (A Picard's Theorem).** Let $J = [x_0 - \varepsilon, x_0 + \varepsilon]$ and $S = J \times E^1$. Let $f : S \to E^1$ be continuous and suppose that $f$ satisfies the Lipschitz condition

$$|f(x, y_1) - f(x, y_2)| \leq r|y_1 - y_2|$$

for all $y_1, y_2 \in E^1$ and some fixed $r \in E^1$. Then the differential equation

$$y' = f(x, y)$$

has a unique solution through $(x_0, y_0)$ over $J$.

**Proof.** Define $F : C(J, E^1) \to C(J, E^1)$, by letting

$$F(h)(x) = f\left(x, y_0 + \int_{x_0}^{x} h(t)dt\right),$$

for every $x \in J$. Then, letting $\delta > r$,

$$d_w(F(h), F(g)) = \sup_x e^{-\delta|x-x_0|} \left| f\left(x, y_0 + \int_{x_0}^x h(t)dt\right) - f\left(x, y_0 + \int_{x_0}^x g(t)dt\right) \right|$$

$$\leq \sup_x e^{-\delta|x-x_0|} r \left| \int_{x_0}^x (h(t) - g(t))dt \right|$$

$$\leq \sup_x r \int_{x_0}^x e^{-\delta|x-x_0|} |h(t) - g(t)| dt$$

$$= r \sup_x \int_{x_0}^x e^{-\delta|x-t|} (e^{-\delta|t-x_0|} |h(t) - g(t)|) dt$$

$$\leq r \sup_x \int_{x_0}^x e^{-\delta|x-t|} d_w(h, g) dt$$

$$= r d_w(h, g) \sup_x \int_{x_0}^x e^{-\delta|x-t|} dt$$

$$= r d_w(h, g) \frac{1}{\delta} \sup_x \int_{x_0}^x e^{-\delta|x-t|} dt$$

$$\leq \frac{r}{\delta} d_w(h, g).$$

Since $\delta > r$, we get that $F$ is a $\frac{r}{\delta}$-contraction. By Theorem 3.27, $F$ has a unique fixed point $u : J \to E^1$; that is, $u(x) = f(x, y_0 + \int_{x_0}^x u(t)dt)$, for every $x \in J$, or letting $v(x) = y_0 + \int_{x_0}^x u(t)dt$, $v'(x) = f(x, v(x))$ for every $x \in J$.

Note that the solution $u : J \to E^1$ is obtained by starting with any guess of it, inasmuch that $u = \lim_n F^n(h)$ for any $h \in C(J, E^1)$; furthermore, Theorem 3.27 gives us the rate of convergence of $F^1(h), F^2(h), \ldots$ to $u$.

**7. Theorem (Peano's Theorem).** Let $J = [x_0 - \varepsilon, x_0 + \varepsilon]$, $S = J \times E^1$ and $f : S \to E^1$ be a continuous and bounded function over $S$. Then the differential equation

$$y' = f(x, y)$$

has a (not necessarily unique) solution through $(x_0, y_0)$ over $J$.

**Proof.** Let $J_0 = [x_0, x_0 + \varepsilon]$ and $I_r = [x_0 + (r-1)\varepsilon/n, x_0 + r\varepsilon/n]$ for $n \in \mathbf{N}$ and integers $I \leq r \leq n$. For each $n$, define a function $y_n : J_0 \to E^1$ by

$$y_n(x) = \begin{cases} y_0, & x_0 \leq x \leq x_0 + \varepsilon/n, \\ y_0 + \int_{x_0}^{x - \varepsilon/n} f(t, y_n(t))dt, & x_0 + (r-1)\varepsilon/n < x \leq x_0 + r\varepsilon/n, \text{ for } r = 2, 3, \ldots, n \end{cases}$$

Note that each $y_n(x)$ is (inductively) well-defined and continuous, inasmuch that, for every $x \in I_r$, $y_n(x)$ is defined in terms of $y_n(t)$ for $t \in I_1 \cup \cdots \cup I_{r-1}$.

Say $|f(x, y)| \leq M$, for every $(x, y) \in S$. If $x > x_0 + \varepsilon/n$ then

$$|y_n(x) - y_0| \leq \int_{x_0}^{x - \varepsilon|n} |f(t, y_n(t))| dt \leq M|x - \varepsilon/n - x_0| < M\varepsilon.$$

If $x_0 \leq x \leq x_0 + \varepsilon/n$ then $|y_n(x) - y_0| = 0$. So, each $\{y_n(x)|n \in \mathbf{N}\}^- \subset [-M\varepsilon, M\varepsilon]$ is compact. Likewise,

$$|y_n(x) - y_n(w)| \leq M|x - w|, \text{ for all } x, w \in J_0.$$

So, $\{y_n|n \in \mathbf{N}\}$ is equicontinuous. Therefore, by Theorem 5, there exist $u \in C(J_0, E^1)$ and a subsequence $\{y_{n_k}\}$ of $\{y_n\}$ such that $\lim_k y_{n_k} = u$.

Finally, we will show that $u$ is a solution of $y' = f(x, y)$: Rewrite

$$y_{n_k}(x) = y_0 + \int_{x_0}^{x} f(t, y_{n_k}(t)) dt - \int_{x - \varepsilon/n_k}^{x} f(t, y_{n_k}(t)) dt$$

and note that

$$\lim_k \int_{x_0}^{x} f(t, y_{n_k}(t)) dt = \int_{x_0}^{x} f(t, u(t)) dt,$$

$$0 \leq \lim_k \left| \int_{x - \varepsilon/n_k}^{x} f(t, y_{n_k}(t)) dt \right| \leq \int_{x - \varepsilon/n_k}^{x} M \, dt = \lim_k M\varepsilon/n_k = 0.$$

It follows that, for every $x \in J_0$,

$$u(x) = y_0 + \int_{x_0}^{x} f(t, u(t)) dt \text{ or } u'(x) = f(x, u(x)) \text{ with } u(x_0) = y_0.$$

Since the preceding argument applies equally well to $[x_0 - \varepsilon, x_0]$, we get another solution $v \in C([x_0 - \varepsilon, x_0], E^1)$ of $y' = f(x, y)$ with $v(x_0) = y_0$. The fact that $u(x_0) = y_0 = v(x_0)$ and $u'(x_0) = f(x_0, y_0) = v'(x_0)$ allows us to "glue" the functions $v$ and $u$, thus obtaining a solution $\gamma : J \to E^1$ of $y' = f(x, y)$ with $\gamma(x_0) = y_0$.

There is really no hope for uniqueness in Peano's Theorem: The equation $y' = 3y^{2/3}$ has the solutions $y(x) = 0$ and $y(x) = x^3$ passing through the point $(0, 0)$ over any interval $[-a, a]$.

**8. Theorem.** For any closed interval $[a, b]$ there exist continuous functions $f : [a, b] \to E^1$ which are nowhere differentiable (*i.e.* $f$ does not have a derivative at any point of $[a, b]$). Indeed,

$$K_{ab} = \{f \in C([a, b], E^1)|f \text{ is nowhere differentiable}\}$$

is dense in $C([a, b], E^1)$.

**Proof.** For $m = 1, 2, \ldots$, consider the condition $S(m) : \left| \frac{f(x_0 + h) - f(x_0)}{h} \right| \leq m$, for some $a \leq x_0 \leq b$ and $0 < |h| < \frac{1}{m}$ such that $a \leq x_0 + h \leq b$ and let

$$A_m = \{f \in C([a,b], E^1) | f \text{ satisfies } S(m)\},$$

$$\tilde{A}_m = C([a,b], E^1) - A_m.$$

Clearly, we only need to show that each $\tilde{A}_m$ is a dense open subset of $C([a,b], E^1)$ and apply the Baire Category Theorem (see ex. 3.19) (note that $K_{ab} = \bigcap_{m=1}^{\infty} \tilde{A}_m$, since $a \le x_0 \le b$ and $|f'(x_0)| \le m$ easily imply that $f$ is contained in some $A_j$, for sufficiently large $j$).

First, $\tilde{A}_m$ is open: Suppose not. Then, there exists $f \in \tilde{A}_m$ and $\{f_i\} \subset A_m$ such that $\lim_i f_i = f$. For $i \in \mathbf{N}$, let $a \le x_i \le b$ such that $\left| \frac{f_i(x_i+h)-f_i(x_i)}{h} \right| \le m$, for every $0 < |h| < \frac{1}{m}$. Pick subsequence $\{x_{i_k}\}$ of $\{x_i\}$ and $a \le x_0 \le b$ such that $\lim_k x_{i_k} = x_0$ (compactness!). It follows that

$$\left| \frac{f(x_0+h)-f(x_0)}{h} \right| = \lim_k \left| \frac{f_{i_k}(x_{i_k}+h)-f_{i_k}(x_{i_k})}{h} \right| \le m,$$

for every $0 \le |h| \le \frac{1}{m}$. That is, it follows that $f \in A_m$, a contradiction.

Finally, $\tilde{A}_m$ is dense: Let $f \in A_m$ and $\varepsilon > 0$. We wish to show that there exists $g \in \tilde{A}_m \cap B(f, \varepsilon)$. Clearly it suffices to show that there exists $g \in B(f, \varepsilon)$ such that whenever $g$ has a derivative $g'(x)$ at $x$ then $|g'(x)| > m$. This is really quite easy to do, even though the details are tedious. Descriptively, from the proof of Lemma l(d), we get that the center $f$ of the ball $B(f, \varepsilon)$ is covered by rectangular subsets $R_1, \ldots, R_n$ with compact base and open height such that $R_i \cap R_{i+1} \neq 0$ for

$i = l, 2, \ldots, n-1$. Choose $y_i \in R_i \cap R_{i+1}$ for $i = l, 2, \ldots, n$, and let $y_0 = f(a)$, $y_n = f(b)$. In each rectangle $R_i$ construct a saw-tooth function $s_i$ with teeth slim enough that $|s_i'(x)| > m$ whenever it exists. Glue the functions $s_1, \ldots, s_n$ thus obtaining a function $g \in \tilde{A}_m \cap B(f, \varepsilon)$.

## 4.3   Approximation

Certainly, after studying either Taylor Series or Fourier Series, the reader is fully aware of the great usefulness of "approximating" certain difficult, but valuable, functions by simpler ones—generally, polynomials (for Fourier or Taylor Series, the polynomials are, of course, the partial sums). It is therefore imperative that one have a good understanding of "approximation". For this, we give two theorems: The first is a *constructive result* and the second is an *existence result*. (There are many other constructive results.)

**9. Lemma.** There exists a sequence $\{q_n(x)\}$ of polynomials, with real coefficients, which converges uniformly on $[0, 1]$ to the function $\varphi(x) = \sqrt{x}$.

**Proof.** Note that the Maclaurin series for $\sqrt{x} = 1 + \binom{0.5}{1}(x - 1) + \binom{0.5}{2}(x - 1)^2 + \binom{0.5}{3}(x - 1)^3 + \cdots$, where $\binom{r}{n} = \frac{r(r-1)\cdots(r-n+1)}{n!}$, for any real number $r$, converges for $0 \le x \le 1$ (this is easily seen by using integral remainders). So, letting $q_n(x)$ be the $n^{th}$-partial sum of this series, we then get that $\{q_n(x)\}$ converges uniformly on $[0, 1]$ to $\varphi(x) = \sqrt{x}$.

**10. Corollary.** For any real numbers $a < b$, there exists a sequence $\{p_n(x)\}$ of polynomials, with real coefficients, which converges uniformly on $[a, b]$ to the function $\psi(x) = |x|$.

**Proof.** Note that the polynomials $p_n(x) = b q_n\left(\frac{|x|^2}{b^2}\right)$ converge uniformly on $[a, b]$ to $b\sqrt{\frac{|x|^2}{b^2}} = |x|$.

For any space $X$ and Euclidean space $E^n$, call a subset $A$ of $C(X, E^n)$ an *algebra* provided that $f, g \in A$ and $\lambda \in E^1$ imply that $f + g \in A$, $fg \in A$ and $\lambda \in A$, where, for each $x \in X$,

$$(f + g)(x) = f(x) + g(x),$$

$$fg(x) = f(x)g(x),$$

$$(af)(x) = af(x).$$

An algebra $A$ in $C(X, E^n)$ is said to *distinguish points of $X$*, provided that, for every $x, y \in X$ with $x \ne y$, there exists $f_{xy} \in A$ such that $f_{xy}(x) \ne f_{xy}(y)$.

For each $\lambda \in E^n$, let $c_\lambda : X \to E^n$ denote the constant function defined by $c_\lambda(x) = \lambda$, for every $x \in X$.

**11. Theorem. (Stone-Weierstrass Approximation.)** Let $X$ be a compact metric space and $A$ an algebra in $C(X, E^1)$ such that (the constant function) $c_I \in A$ and $A$ distinguishes points of $X$. Then $A^- = C(X, E^1)$.

**Proof.** Without loss of generality, we assume that $A$ is closed (since it is trivial to check that the closure $A^-$ of an algebra $A$, is also an algebra—see ex. 3) and we proceed to show that $A = C(X, E^1)$.

First, $f \in A$ implies $|f| \in A$: Say $f(X) \subset [a, b]$, since $X$ is compact. Let $\varepsilon > 0$ and choose a polynomial $p(t) = a_0 + a_1 t + \cdots + a_n t^n$ such that

$$||t| - p(t)| < \varepsilon, \text{ for every } t \in [a, b].$$

(This can be done, by Corollary 10.) Then, letting $p(f) = a_0 + a_1 f + \cdots + a_n (f)^n$, with $(f)^{k+1}(x) = (f(x))^{k+1}$, for $k \in \mathbf{N}$ and $x \in X$, it follows that

$$d_s(|f|, p(f)) < \varepsilon,$$

with $p(f) \in A$. Therefore, since $\varepsilon$ is arbitrary, $f \in A$ implies $|f| \in A^- = A$.

Next, $f_1, f_2, \ldots, f_n \in A$ implies $\min(f_1, \ldots, f_n) \in A$ and $\max(f_1, \ldots, f_n) \in A$: (note that $\min(f_1, \ldots, f_n)(x) = \min(f_1(x), \ldots, f_n(x))$, $\max(f_1, \ldots, f_n)(x) = \max(f_1(x), \ldots, f_n(x))$). Since, for example, $\min(f_1, \ldots, f_n) = \min(\min(f_1, \ldots, f_{n-1}), f_n)$, it suffices to verify our claim for two functions $f, g \in A$. Therefore it suffices to check that (see ex. 4)

$$\min(f, g) = \frac{1}{2}(f + g) - \frac{1}{2}|f - g|,$$

$$\max(f, g) = \frac{1}{2}(f + g) + \frac{1}{2}|f - g|$$

and note that $|f - g| \in A^- = A$.

We are now ready for the final assault, which will be done in two stages.

(i) $a \in X$, $f \in C(X, E^1)$ and $\varepsilon > 0$ implies that there exists $f_a \in A$ such that $f_a(a) = f(a)$ and $f_a(x) < f(x) + \varepsilon$ for every $x \in X$: Since $A$ is an algebra, for every $b \in X$, there exists $f_{ab} \in A$ such that $f_{ab}(a) = f(a)$ and $f_{ab}(b) = f(b)$ (for example, let $h \in A$ such that $h(a) \neq h(b)$ and let $f_{ab}(x) = f(a)\frac{h(x)-h(b)}{h(b)-h(a)} + f(b)\frac{h(x)-h(a)}{h(b)-h(a)}$, for every $x \in X$). By continuity of the functions $f$ and $f_{ab}$, there exists a neighborhood $N_b$ of $b$ such that $f(N_b) \cup f_{ab}(N_b) \subset] f(b) - \varepsilon/2, f(b) + \varepsilon/2[$, from which it follows that

$$z \in N_b \text{ implies } f_{ab}(z) < f(z) + \varepsilon.$$

By compactness of $X$, let $N_{b_1}, \ldots, N_{b_n}$ cover $X$. Then it is easily seen that the function

$$f_a = \min(f_{ab_1}, \ldots, f_{ab_n})$$

satisfies all requirements.

(ii) $f \in C(X, E^1)$, $\varepsilon > 0$ implies that there exists $\tilde{f} \in A$ such that $d_s(f, \tilde{f}) < \varepsilon$: For each $a \in X$, choose a function $f_a$ satisfying (i) above. Then, there exists a neighborhood $V_a$ of $a$ such that

$$z \in V_a \text{ implies } f_a(z) > f(z) - \varepsilon \text{ (better yet, } \varepsilon > f(z) - f_a(z)).$$

By compactness of $X$, let $N_{a_1}, \ldots, N_{a_k}$ cover $X$. It follows that, letting $\tilde{f} = \max(f_{a_1}, \ldots, f_{a_n})$,

$$z \in X \text{ implies } f(z) - \varepsilon < \tilde{f}(z) < f(z) + \varepsilon;$$

that is, $d_s(f, \tilde{f}) < \varepsilon$. Since $\varepsilon$ is arbitrary, and $\tilde{f} \in A$, we get that $f \in A$. This shows that $A = A^- = C(X, E^1)$.

Theorem 11 can be generalized to other Euclidean spaces, but one must be careful (see ex. 17). Theorem 11, even though non-constructive, has a great virtue: It alerts us to various possible collections of functions which suffice to approximate any given $f \in C(X, E^1)$. Indeed, for every $S \subset C(X, E^1)$, let

$$A(S) = \bigcap \{K | K \text{ is a closed subalgebra of } C(X, E^1), S \subset K\};$$

it is straightforward that $A(S)$ is a closed algebra; $A(S)$ is called the *closed algebra generated by* $S$ (in $C(X, E^1)$, of course). It then follows that, for example,

  (i) $A(\{1, x\}) = C([a, b], E^1)$,
 (ii) $A(\{1, x^2\}) = C([0, t], E^1)$,
(iii) $A(\{1, x^2\}) \neq C([-1, 1], E^1)$, because it does not distinguish $-1$ from $1$,
(iv) $A(\{1, \cos x\}) = C([0, \pi], E^1)$,
 (v) $A(\{1, 1 + \cos \alpha\}) = C([0, 2\pi], E^1)$.

## 4.4  Function-Space Functions

By now, we are certain that the reader will agree that function spaces are extremely important and very difficult to handle; that the *pc* topology is rather useless, while the *uc* topology is very useful; and finally, that the *uc* and *co* topologies are closely related (cf. Lemma l(d) and ex. 18). The following three constructions and subsequent results not only simplify the *co* and *uc* topologies but also provide us with crucial tools for some of the work ahead.

*Composite Function.* For any spaces $X$, $Y$, $Z$ define $T : Y^X \times Z^Y \to Z^X$, by letting

$$T(f, g) = g \circ f, \text{ for every } (f, g) \in Y^X \times Z^Y.$$

(The function T is called the *composition function.*)

*Evaluation Function.* For any spaces $X$, $Y$, define $\xi : Y^X \times X \to Y$ by letting

$$\xi(f, x) = f(x), \text{ for every } (f, x) \in Y^X \times X.$$

(The function $\xi$ is called the *evaluation function.*)

*Associated Functions.* For any spaces $X$, $Y$, $Z$ note that any function $f : X \times Y \to Z$ generates a function $\hat{f} : X \to Z^Y$ defined by

$$(\hat{f}(x))(y) = f(x, y)$$

for every $(x, y) \in X \times Y$; conversely, any $\hat{f}$ generates an $f$, by the same equation above. The functions $f$ and $\hat{f}$ are called *associated functions.*

**12. Lemma.** The composition function $T : Y^X \times Z^Y \to Z^X$ is continuous with respect to the *co* topology on all spaces, whenever $Y$ is locally compact Hausdotff.

**Proof.** By Theorem 1.14(vi), it suffices to prove that $T^{-1}\langle K, U \rangle$ is open, for each $\langle K, U \rangle \in S_{co}$. But observe that, for each $(f, g) \in T^{-1}\langle K, U \rangle$, we get that

$$
\begin{array}{ccccc}
X & \xrightarrow{f} & Y & \xrightarrow{g} & Z \\
\uparrow U & & \uparrow U & & \uparrow U \\
K & & f(k) & & U
\end{array}
$$

with $f(K)$ compact. Since $g$ is continuous, $g(f(K)) \subset U$ and $Y$ is locally compact Hausdorff, there exists open cover $\mathcal{V}$ of $f(K)$ such that $V^-$ is compact and $g(V^-) \subset U$, for every $V \in \mathcal{V}$ (see Lemma 3.19). Let $\{V_1, \ldots, V_n\}$ be a finite subcollection of $\mathcal{V}$ which covers (the compact space) $f(K)$. Then $W = \bigcup_{i=1}^n V_i$ is open in $Y$, $W \supset f(K)$, $W^- = \bigcup_{i=1}^n V_i^-$ is compact and $g(W^-) \supset U$. It follows that

$$\langle K, W \rangle \times \langle W^-, U \rangle \subset T^{-1}(\langle K, U \rangle)$$

is a neighborhood of $(f, g)$, which shows that $T^{-1}\langle K, U \rangle$ is open. This completes the proof.

**13. Corollary.** The evaluation function $\xi : Y^X \times X \to Y$ is continuous with respect to *co* topology, whenever $X$ is locally compact.

**Proof.** First note that, with $1 = \{0\}$, $X^1 \simeq X$. Since $\xi$ is the composition of the maps

$$Y^X \times X^1 \xrightarrow{\ j\ } X^1 \times Y^X \xrightarrow{\ T\ } Y^1 \simeq Y,$$

with $j(f, x) = (x, f)$, for every $(f, x) \in Y^X \times X^1$ it follows from Lemma 12 that $\xi$ is continuous (obviously, $j$ is continuous).

**14. Theorem.** For spaces $X, Y, Z$, with $Y$ locally compact Hausdorff, $f : X \times Y \to Z$ is continuous iff its associated function $\hat{f} : X \to Z^Y$ is continuous.

**Proof.** First, the "if" part: Let $\hat{f} : X \to Z^Y$ be continuous and note that $f = \xi \circ (\hat{f} \times i_Y)$,

$$X \times Y \xrightarrow{\ \hat{f} \times i_Y\ } Z^Y \times Y \xrightarrow{\ \xi\ } Z,$$

with $i_Y : Y \to Y$ being the identity function. Since all these functions are continuous (see Lemma 2.7), it follows that $f$ is continuous.

Now, the "only if" part: Let $f : X \times X \to Z$ be continuous. To show that $\hat{f} : X \to Z^Y$ is continuous, we pick $\langle K, V \rangle \in S_{co}$ for $Z^Y$ and show that $f^{-1}(\langle K, V \rangle)$ is open:

Let $x \in f^{-1}(\langle K, V \rangle)$. Then, for every $y \in K$,

$$f(x, y) = (\hat{f}(x))(y) \in V.$$

Since $f$ continuous, for every $y \in K$, there exists an open neighborhood $N_{xy} \times N_y$ of $(x, y)$ in $X \times X$ such that

$$f(N_{xy} \times N_y) \subset V.$$

Since $K$ is compact, let $N_{y_i}, \ldots, N_{y_n}$ cover $K$. Then, letting $N_x = \bigcap_{i=1}^n N_{xy_i}$ and $N_K = \bigcup_{i=1}^n N_{y_i}$, it follows that

$$\{x\} \times K \subset N_x \times N_K, N_x \times N_K \text{ is open,}$$

$$(\hat{f}(N_x))(K) \subset (\hat{f}(N_x))(N_K) = f(N_x \times N_K) \subset V.$$

Therefore $N_x$ is a neighborhood of $x$ such that

$$N_x \subset \hat{f}^{-1}(\langle K, V \rangle),$$

which completes the proof.

# Chapter 4.  Exercises

1. A sequence $\{f_n\}$ of functions $f_n : X \to E^1$ is said to be *decreasing (increasing)* provided that $f_n(x) \geq f_{n+1}(x)$ $(f_n(x) \leq f_{n+1}(x))$, for every $x \in X$ and $n \in \mathbf{N}$.

   (i) If $X$ is a space, each $f_n$ is continuous, $\{f_n\}$ is decreasing (or increasing) and, for some $p \in X$, $\lim\limits_n f_n(p) = 0$, show that for every $\varepsilon > 0$, there exist $n(\varepsilon)$ and neighborhood $N_p$ of $p$ such that

   $$f_n(x) < \varepsilon, \text{ for every } n > n(\varepsilon) \text{ and } x \in N_p.$$

   (ii) *Theorem of Dini.* Let $(X, \rho)$ be a compact metric space and $\{f_n\}$ a decreasing sequence of functions such that $\{f_n\}$ converges pointwise to the *continuous* function $f : X \to E^1$. Then $\{f_n\}$ converges to $f$ uniformly.

2. Let $Y^X$ and $Z^Y$ be function spaces, such that the sup metric makes sense in $Z^Y$ (for example, $Y$ compact and $(Z, \rho)$ metric). Prove that, for every $h \in Y^X$ and $f, g \in Z^Y$,

   $$\rho_s(f, g) \geq \rho_s(f \circ h, g \circ h).$$

3. For any compact space $X$, let $A$ be an algebra in $C(X, E^n)$ with the sup metric. Show that $A^-$ is also an algebra. (Hint: Given $f, g \in A^-$ and $\lambda \in E^1$, let $\{f_n\} \subset A$ and $\{g_n\} \subset A$ such that $\lim\limits_n d_s(f_n, f) = 0$ and $\lim\limits_n d_s(g_n, g) = 0$. Now, show that $\lim\limits_n d_s(f_n + g_n, f + g) = 0$, $\lim\limits_n d_s(f_n g_n, fg) = 0$ and $\lim\limits_n d_s(\lambda f_n, \lambda f) = 0$, keeping in mind that the metric $d$ is the *Euclidean metric*.)

4. Let $X$ be compact Hausdorff and $f, g \in C(X, E^1)$. Show that

   (i) $\min(f, g) = \frac{1}{2}(f + g) - \frac{1}{2}|f - g|$,

   (ii) $\max(f, g) = \frac{1}{2}(f + g) + \frac{1}{2}|f - g|$.

   (Hint: For example, for (i) consider the two cases: Case 1. $\min(f, g)(x) = f(x)$; Case 2. $\min(f, g)(x) = g(x)$. Check that the formula (i) works in either case.)

5. Let $C^\infty$ be the set of all infinite sequences $\{x_n\}$ of complex numbers such that $\Sigma_n |x_n|^2 < \infty$. Let $\rho_1(\bar{x}, \bar{y}) = \sup\{|x_n - y_n| : n \in \mathbf{N}\}$ and $\rho_2(\bar{x}, \bar{y}) = \sqrt{\sum_{n=1}^{\infty} |x_n - y_n|^2}$, for every $\bar{x}, \bar{y} \in C^\infty$. Show that

   (a) $\rho_1$ is a complete metric on $C^\infty$.

   (b) $(C^\infty, \rho_2)$ is a complete metric space (this space is known as the *Hilbert space*).

6. For $i = 1, 2, \ldots, n$, let $J_i$ be a closed interval and let $C_n = \Pi_{i=1}^n C(J_i, E^1)$, with the metric $d((f_i), (g_i)) = \max\limits_i d_w(f_i, g_i)$; (cf. Lemma 2(ii)). Show that $(C_n, d)$ is a complete metric space.

7. Let $f(x) = x + 1$, for every $x \in E^1$. Is $\{f^n | n \in \mathbf{N}\}$ an equicontinuous family?

8. Let $(X, d)$ be a metric space and $f : X \to X$ an $\alpha$-contraction. Show that

   (a) $\Lambda(f) = \{f^n | n = 1, 2, \ldots\}$ is equicontinuous.

   (b) $\Gamma(f) = \Lambda(f)^- \subset (X^X, d_s)$ is equicontinuous.

(c) If $X$ is compact, $\Gamma(f)$ is compact.

9. Let $f(x) = x^2$, for $x \in I$. Show that $\{f^n | n = 1, 2, \ldots\}$ is not equicontinuous.

10. Let $(Y, d)$ be a metric space and $\mathcal{F}$ be a compact subset of $(Y^X, d_s)$. Show that $\mathcal{F}$ is equicontinuous. (Hint: Suppose not. Then there exists $p \in X$ and $\varepsilon > 0$ such that, for each $B(p, \frac{1}{n})$, there exists $f_n \in \mathcal{F}$ such that $f_n(B(p, \frac{1}{n})) \not\subset B(f_n(p), \varepsilon)$. Therefore, there exists $\{x_n\} \subset X$ such that $\lim_n x_n = p$ and $d(f_n, (x_n), f_n(p)) \geq \varepsilon$. Since $\mathcal{F}$ is compact the sequence $\{f^n\}$ has a cluster point $f \in \mathcal{F}$. Observing that

$$\varepsilon \leq d(f_n(x_n), f_n(p)) \leq d(f_n(x_n), f(x_n)) + d(f(x_n), f(p)) + d(f(p), f_n(p)),$$
$$\leq d_s(f_n, f) + d(f(x_n), f(p)) + d_s(f, f_n),$$

complete the proof.)

Also show that $\mathcal{F}$ is closed in $(Y^X, d_s)$ and that $Z_X = \{f(x) | f \in \mathcal{F}\}$ is compact. (Note that $\mathcal{F}$ is compact. Is $w_x : Y^X \to X$, defined by $w_x(f) = f(x)$, a continuous function?)

11. *Linear Integral Equations.* In physics, equations of the form

$$(1) \quad x(s) = \lambda \int_a^b K(s,t) x(t) dt + f(s)$$

appear frequently. This particular one is a simple Fredholm's equation of the second kind and appears in the study of *small oscillations of elastic systems.* Prove that if $K$ is defined and continuous on the rectangle $R = \{(s,t) | a \leq s, t \leq b\}$, $f$ is defined and continuous on the interval $J = [a, b]$ and $|\lambda| < \frac{1}{(b-a)M}$ with $M = \sup\{K(s,t) | (s,t) \in R\}$, then the above equation has a unique solution. (Hint: Consider the space $C(J, J)$ and define $F : C(J, J) \to (J, J)$ by letting

$$F(g)(s) = \lambda \int_a^b K(s,t) g(t) dt + f(s),$$

for every $s \in J$.) Show that $F$ is a contraction. (Why must we have that $f$ is continuous?) Show that (1) has a unique solution $x = x(t)$.

12. *Nonlinear Integral Equations.* In physics, one also encounters equations of the form

$$(2) \quad x(s) = \lambda \int_a^b K(s,t,x(t)) dt + \beta(s)$$

with $K$ defined and continuous on a parallelepiped $P = \{(s,t,z) | a \leq s, t \leq b, -H \leq z \leq H\}$ and the function $x(t)$ defined on $J = [a, b]$. Prove that if the function $K$ satisfies the Lipschitz condition

$$|K(s,t,x_1) - K(s,t,x_2)| \leq L|x_1 - x_2|$$

for all $x_1, x_2 \in [-H, H]$ and $|\lambda| < \frac{1}{L - (b-a)}$, then the above equation has a unique solution. (Hint: Consider the space $C(J, [-H, H])$ and define $F$ in $C(J, [-H, H])$ by letting

$$F(g)(s) = \lambda \int_a^b K(s,t,g(t)) dt + \beta(s)$$

for every $s \in J$.) Show that $F$ is a contraction. Show that (2) has a unique solution $u = x(t)$.

13. *Volterra Equations.* Let $J = [a, b] \subset E^1$, $a \geq 0$, $\lambda > 0$ and $\varphi : C(J, E^1) \to C(J, E^1)$ be defined by

$$\varphi(f)(x) = \lambda \int_a^x K(x, y) f(y) dy + \mu(x), \text{ for every } x \in J.$$

For $g, h \in C(J, E^1)$, let $M = \sup_{x,y} |K(x, y)|$ and $t = \sup_x |g(x) - h(x)|$. Show that, letting $\rho$ be the Euclidean metric,

(a) $\rho_s(\varphi(g), \varphi(h)) \leq \lambda M t (b - a)$.
(b) $\rho_s(\varphi^m(g), \varphi^m(h)) \leq t(\lambda M(b - a))^m / m!$ (Use induction!)
(c) There exists $n$ such that $(\lambda M(b - a))^n / n! < 1/t$.
(d) $\varphi^n$ is a contraction.
(e) $\varphi$ has a unique fixed point (cf. ex. 3.16).

14. *Infinite Systems of Linear Equations.* Let

(1) $y_m = \Sigma_n a_{mn} x_n + c_m$, for $m = 1, 2, \ldots$.

This can be put in vector notation $\bar{y} = A\bar{x} + \bar{c}$, where $\bar{x} = (x_1, x_2, \ldots)$, $\bar{y} = (y_1, y_2, \ldots)$, $\bar{c} = (c_1, c_2, \ldots)$ are written in column form and $A = (a_{ij})$ is an infinite matrix. Show that (see ex. 5)

(a) If $\sup_m \Sigma_n |a_{mn}| < |$, then we can define a function $f : (C^\infty, \rho_1) \to (C^\infty, \rho_1)$, by letting $f(\bar{x}) = A\bar{x} + \bar{c}$; furthermore, this function is a contraction. (This implies that $f$ has a unique fixed point, because of ex. 5(a), and (1) has a unique solution.)

(b) If $\sup_m \Sigma_n |a_{mn}|^2 < 1$, then we can define a function $f : (C^\infty, \rho_2) \to (C^\infty, \rho_2)$, by letting $f(\bar{x}) = A\bar{x} + \bar{c}$; furthermore this function is a contraction. (This implies that $f$ has a unique fixed point, because of ex. 5(b), and (1) has a unique solution.)

15. *Finite Systems of O.D.E.'s of First Order.* Let

(3) $y_i'(x) = \varphi_i(x, y_1(x), \ldots, y_n(x))$, $i = 1, 2, \ldots$, be a system of differential equations with initial conditions $x = x_0$ and $y_i(x_0) = y_{i0}$, $i = 1, 2, \ldots$. Assume the functions $\varphi_i$ are continuous in some cube $B(x_0, \varepsilon) \times \Pi_{j=1}^n B(y_j, \varepsilon) \subset E^{n+1}$ and satisfy the Lipschitz condition.

(4) $|\varphi_i(x, \bar{y}) - \varphi_i(x, \bar{z})| \leq M \max_i |y_i - z_i|$, for $i = 1, 2, \ldots$, where $\bar{y}, \bar{z} \in E^n$.

Show that

(a) To solve (3) and (4) is equivalent to solving (4) and

(5) $y_i(x) = y_{i0} + \int_{x_0}^x \varphi_i(t, y_1(t), \ldots, y_n(t)) dt, i = 1, 2, \ldots$.

(b) The function $A : (C_n, d) \to (C_n, d)$ (cf. ex. 6), defined by the $n$-tuple

$$A(g_1, \ldots, g_n)(x) = (y_{i0} + \int_{x_0}^x \varphi_i(t, g_1(t), \ldots, g_n(t)) dt)_i,$$

is a contraction. Hint: Note that, taking $\delta > M$ and the proof of

Theorem 6,

$$d(A(g_1, \ldots, g_n), A(h_1, \ldots, h_n))$$

$$\leq \max_i \sup_x e^{-\delta|x-x_0|} \int_{x_0}^x |\varphi_i(t, g_1(t), \ldots, g_n(t))$$
$$- \varphi_i(t, h_1(t), \ldots, h_n(t))| dt$$

$$\leq \max_i \sup_x e^{-\delta|x-x_0|} \int_{x_0}^x M \max_i |g_i(t) - h_i(t)| dt$$

$$\leq \max_i \sup_x \int_{x_0}^x M e^{-\delta|x-t|} \max_i e^{-\delta|t-x_0|} |g_i(t) - h_i(t)| dt$$

$$\leq \max_i \sup_x \int_{x_0}^x M e^{-\delta|x-t|} \max_i d_w(g_i, h_i) dt$$

$$= \max_i \frac{M}{\delta} d_w(g_i, h_i) \sup_x \int_{x_0}^x e^{-\delta|x-t|} \delta \, dt$$

$$\leq \frac{M}{\delta} \max_i d_w(g_i, h_i) = \frac{M}{\delta} d((g_1, \ldots, g_n), (h_1, \ldots, h_n))$$

(c) (3) and (4) have a unique solution.

16. *O.D.E.'s of Order $n$.* Consider the differential equation

    (6) $y^{(n)} = F(y, y', \ldots, y^{(n-1)}, x)$

    and show that

    (a) Solving (6) is equivalent to solving the finite system of o.d.e.'s of first order

    (7) $\begin{cases} y = y_1, y' = y_2, \ldots, y^{n-1)} = y_n \\ y_k' = y_{k+1}, k = 1, \ldots, n-1 \\ y_n' = F(y_1, y_2, \ldots, y_n, x) \end{cases}$

    (b) Use ex. 15, to obtain conditions that guarantee existence and uniqueness of solutions for (7) and the initial conditions $y_i(x_0) = y_{i0}, i = 1, \ldots, n$; equivalently for (6) and the initial conditions $y^{(i)}(x_0) = y_{i0}, i = 1, \ldots, n$.

17. Show that Theorem 11 becomes false if we replace $E^1$ by $E^2$ in its statement. (Hint: Let $X = [B((0,0), 1)]^-$ and $A$ be the algebra of all polynomials in $z = x + iy$, with real coefficients. Show that the (constant function) $c_i \notin A^-$, even though $c_1 \in A$ and $A$ distinguishes points by $p(z) = z$.)

    Now, show that Theorem 11 becomes valid with $E^1$ replaced by $E^2$ if we add the condition that $f \in A$ implies the conjugate function $\bar{f} \in A$ (*i.e.* if $f(z) = f_1(z) + if_2(z)$ then $\bar{f}(z) = f_1(z) - if_2(z)$).

18. *Broadening the Ascoli-Arzela Theorem.* Let $X$ be a metric space for which there exists a sequence $\{C_n\}$ of compact subspaces with $X = \bigcup_{n=1}^\infty C_n$ (*i.e.* $X$ is $\sigma$-compact) and let $(Y, d)$ be any metric space. For each $n \in \mathbf{N}$ and $f, g \in Y^X$, let

$$d_n(f, g) = \sup\{d(f(x)), g(x)) | x \in C_n\}$$

and let

$$d_\Sigma(f, g) = \Sigma_{n=1}^\infty \frac{1}{2^n} \frac{d_n(f, g)}{1 + d_n(f, g)}.$$

(i) Show that $d_\Sigma$ is a metric on $Y^X$,

(ii) Letting $K_n = \{f|C_n : f \in Y^X\}$, for each $n$, show that $(K_n, d_\Sigma|K_n \times K_n)$ is a metric subspace of $(C(C_n, Y), d_n)$. (Be careful!) These two spaces are generally not identical, since there may exist a continuous function $g : C_n \to Y$ such that *no continuous* $\tilde{g} : X \to Y$ satisfies $\tilde{g}|C_n = g$.)

(iii) Show that a family $\mathcal{F} \subset (Y^X, d_\Sigma)$ is compact iff each $\mathcal{F}_n = \{f|C_n : f \in \mathcal{F}\}$ is compact. Also show that each $\mathcal{F}_n$ is compact iff $\mathcal{F}_n$ is equicontinuous and $Z_{xn} = \{f(x)|f \in \mathcal{F}\}^-$ is compact, for each $x \in C_n$ (Hint: Immediate from (ii) and Theorem 5.)

# Chapter 5

# Topological Groups

When dealing with integrals, areas and volumes, and continuous functions we use the following principles with no second thoughts:

(i) If $f, g : E^1 \to E^1$ are continuous then so are $f + g$, $f - g$ and $fg$.

(ii) In $E^3$ we can rotate and translate geometric figures without changing their volume, area or length.

And yet we should have many second thoughts about this. After all, letting

$$\mathcal{B} = \{[a, b[ \, | \, a, b \in E^1, a < b\}$$

it is easy to see that $\mathcal{B}$ is a base for a topology $\tau_h$ on $E^1$ (see ex. 1.3). It is equally easy to see that

(iii) The identity function $j : (E^1, \tau_h) \to (E^1, \tau_h)$ and the constant functions $c_a : (E^1, \tau_h) \to (E^1, \tau_h)$ are continuous. However the functions $c_0 - j = -j$ and $k$, defined by $k(x) = j(x)j(x) = x^2$, are not continuous.

At the end of Section 2.2 we pointed out that, with respect to the metric $d((x_1, x_2), (y_1, y_2)) = |x_1 - y_1| + |x_2 - y_2|$ on $E^2$, the length of the segment from $(0, 0)$ to $(1, 1)$ is 2. However, it is immediate that the length of its rotation, the segment from $(0, 0)$ to $(0, \sqrt{2})$, is $\sqrt{2}$.

At this point, the reader may feel cheated and dejected, or else may ask: Why is it that, with respect to the Euclidean topology the sum, subtraction, product and quotient (whenever the denominator is $\neq 0$) of continuous functions $f, g : E^n \to E^m$ are continuous? Why is it that, with respect to the Euclidean metric on $E^n$, rotation and translation do not affect length, area or volume of geometric figures?

We will now devote our attention to the first question. We will deal with the second question in the exercises (cf. ex. 12).

## 5.1 Elementary Structures

**1. Definition.** A *topological group* is a triple $(G, \square, \tau)$ such that $(G, \square)$ is a group, $(G, \tau)$ is a topological space and

(i) the group operation $\square : G \times G \to G$ is continuous,

(ii) the inversion function $i : G \to G$, defined by $i(x) = x^{-1}$ or $i(x) = -x$, is continuous,

(iii) letting $e$ be the unit element of $G$, $\{e\}$ is a closed subset of $(G, \tau)$.

(There exist mathematicians that do not require the last condition in the definition of a topological group.)

Throughout, we will use either the multiplicative or the additive notation for groups, depending on which seems most convenient. As customary, the juxtaposition $ab$ of two elements $a \in G$, $b \in G$ means *the product of $a$ and $b$ in the group $G$*. The *unit element* of a multiplicative group will generally be denoted by 1 and the *identity element* of an additive group will be denoted by 0.

**2.  Lemma.**  For $n \in \mathbf{N}$, $E^n$ (resp. $E^n - \{0\}$), with the usual coordinatewise addition (resp. multiplication) and the Euclidean topology, is a topological group.

**Proof.**  We will only do the additive case since the other is similar. Clearly, we need only prove that the addition and inversion functions are continuous, since the remaining details should be well known to the reader.

To show that $+ : E^n \times E^n \to E^n$ is continuous, it suffices to check that for every $((a_1, \ldots, a_n), (b_1, \ldots, b_n)) \in E^n \times E^n$ and $B((a_1 + b_1, \ldots, a_n + b_n), \varepsilon)$,

$$B((a_1, \ldots, a_n), \varepsilon/2n) + B((b_1, \ldots, b_n), \varepsilon/2) \subset B((a_1 + b_1, \ldots, a_n + b_n), \varepsilon) :$$

To be precise, we should say

$$+(B((a_1, \ldots, a_n), \varepsilon/2n), B((b_1, \ldots, b_n), \varepsilon/2n)) \subset B((a_1 + b_1, \ldots, a_n + b_n), \varepsilon),$$

but *controlled* imprecision sometimes has its unjust rewards—simplicity, familiarity and convenience.) Simply note that, whenever $|(x_1, \ldots, x_n) - (a_1, \ldots, a_n)| < \varepsilon/2n$ and $|(w_1, \ldots, w_n) - (b_1, \ldots, b_n)| < \varepsilon/2n$, then $|x_i - a_i| < \varepsilon/2n$, $|w_i - b_i| < \varepsilon/2n$ for $i = 1, 2, \ldots, n$; therefore,

$$|(x_1 + w_1, \ldots, x_n + w_n) - (a_1 + b_1, \ldots, a_n + b_n)|$$

$$= \left( \sum |x_i + w_i - a_i - b_i|^2 \right)^{\frac{1}{2}}$$

$$\leq \left( \sum (|x_i - a_i| + |w_i - b_i|)^2 \right)^{\frac{1}{2}}$$

$$< \left( \sum \left( \frac{\varepsilon}{2n} + \frac{\varepsilon}{2n} \right)^2 \right)^{\frac{1}{2}} = \left( \sum_{i=1}^{n} \left( \frac{\varepsilon}{n} \right)^2 \right)^{\frac{1}{2}}$$

$$\leq n \left( \frac{\varepsilon^2}{n^2} \right)^{\frac{1}{2}} = \varepsilon.$$

It is obvious that the inversion function is continuous, since

$$i(B((x_1, \ldots, x_n), \varepsilon)) = B((-x_1, \ldots, -x_n), \varepsilon),$$

for every $(x_1, \ldots, x_n) \in E^n$ and $\varepsilon > 0$.

It is now obvious that sums, products, differences and well-defined quotients of continuous functions $f, g : E^n \to E^m$ are continuous; for example,

$$f + g : E^n \to E^m$$

is the composite of

$$E^n \xrightarrow{(f,g)} E^m \times E^m \xrightarrow{+} E^m,$$

with $(f, g)(x) = (f(x), g(x))$, for every $x \in E^n$. If $h : E^n \to E^m - \{0\}$ then $\frac{1}{h} : E^n \to E^m - \{0\}$ is the composite of

$$E^n \xrightarrow{h} E^m \xrightarrow{i} E^m.$$

Let us recall that, for any group $G$ and $c \in G$, the functions

$$L_c : G \to G, R_c : G \to G,$$

defined by $L_c(x) = cx$, $R_c(x) = xc$, for every $x \in G$, are called *left translations* and *right translations*, respectively. Note that $L_c$ and $R_c$ are actually one-to-one and onto functions whose inverses are $L_{c^{-1}}$, and $R_{c^{-1}}$, respectively. (Note that the geometric translations in $E^n$ are the additive translations. The multiplicative translations in $E^n$ really do no more than "expand" or "contract" the geometric figures.)

Let us also recall that, for any group $G$ and subgroup $H$ of $G$, the collection $G/H = \{aH | a \in G\}$, with the operation

$$(aH)(bH) = abH,$$

is a group iff $H$ is *normal* (*i.e.*, $gHg^{-1} \subset H$, for every $g \in G$) and, in this case, we let

$$G/H = \{aH | a \in G\} = \{Ha | a \in G\},$$

because each $aH = Ha$. We will let $\lambda : G \to G/H$ be the natural homomorphism. Whenever convenient, let $G/H \equiv \frac{G}{H}$.

**3. Lemma.** The following are valid:

(a) If $(G, \square, \tau)$ is a topological group and $H$ is a subgroup of $G$, then $(H, \square, \tau | H)$ is a topological group.

(b) In a topological group $(G, m, \tau)$, the inversion function $i$ and the translations are homeomorphisms.

(c) In a topological group $(G, m, \tau)$, the product of any subset $A$ by any open subset $U$ ($AU = \{au | a \in A, u \in U\}$) is open. In particular, the multiplication $m$ is an open function.

(d) If $(G, m, \tau)$ is a topological group and $H$ is a normal subgroup of $G$, then $\lambda : G \to (G/H,$ quotient topology $\tau_\lambda)$ is open. The singleton $\{H\}$, consisting of the unit element $H$ of $G/H$, is closed with respect to $\tau_\lambda$ iff $H$ is a closed subset of $G$.

(e) If $(G, m, \tau)$ is a topological group and $H$ is a closed normal subgroup of $G$ then $G/H$ with the quotient topology $\tau_\lambda$ is a topological group.

**Proof.** Part (a) is obvious, since restrictions of continuous functions remain continuous, with respect to the subspace topology.

Part (b). Note that the inversion function equals its own inverse function. With respect to translations, it suffices to note that, for every $c \in G$ (including $c^{-1} \in G!$),

$$L_c = m|\{c\} \times G, R_c = m|G \times \{c\}.$$

Part (c). It suffices to note that

$$AU = \bigcup \{L_a(U)|a \in A\}$$

and apply part (b) as well as Corollary 1.16.

Part (d). To show that the quotient map $\lambda : G \to G/H$ is open, pick any open $U \subset G$ and note that

$$\lambda^{-1}(\lambda(U)) = HU$$

is open in $G$. Therefore, $\lambda(U)$ is open because $\lambda$ is a quotient map. To show that $\{H\}$ is closed in $G/H$ iff $H$ is closed in $G$, simply observe that

$$\lambda^{-1}(G/H - \{H\}) = G - H.$$

Therefore, $G - H$ is open in $G$ iff $G/H - \{H\}$ is open in $G/H$, or equivalently, $H$ is closed in $G$ iff $\{H\}$ is closed in $G/H$.

Part (e). Because of part (d), we only need to show that the multiplication and inversion induced by $\lambda$ on $G/H$ are continuous with respect to $\tau_\lambda$. For this, it suffices to check that, the diagrams below are *commutative* (*i.e.* $\lambda \circ m = m' \circ (\lambda \times \lambda), \ldots$)

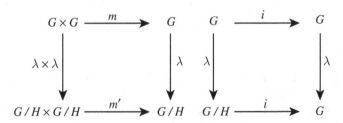

with $\lambda$ and $\lambda \times \lambda$ quotient maps (to show that $\lambda \times \lambda$ is quotient use part (d) and Lemma 2.7 to get that $\lambda \times \lambda$ is open and continuous; then use Lemma 2.14 to get that $\lambda \times \lambda$ is a quotient function).

Let us now recall, for any subset $S$ of a group $(G, m)$,

$$S^{-1} = \{x^{-1} | x \in S\},$$

with $x^{-1}$ denoting the inverse of $x$ in $G$. Furthermore, $S \subset G$ is said to be *symmetric* provided that

$$S = S^{-1}.$$

**4. Lemma.** If $U$ is a neighborhood of $e$ in a topological group $(G, m, \tau)$, then there exists an open symmetric neighborhood $V$ of $e$ such that $VV = VV^{-1} \subset U$. Furthermore, $V^- \subset U$.

**Proof.** By continuity of $m$, there exist open neighborhoods $N$ and $M$ of $e$ such that $NM \subset U$. Let $K = N \cap M$. By Lemma 3(b), $K^{-1}$ is also a neighborhood of $e$. Finally, let

$$V = K \cap K^{-1}.$$

It follows that $V$ is a symmetric neighborhood of $e$ with

$$VV^{-1} = VV \subset KK \subset NM \subset U.$$

Now, we show that $V^- \subset U$: Take any $p \notin U$. Then, by Lemma 3(b), $pV$ is a neighborhood of $p$ which misses $V$. (Say $q \in V \cap pV$. Then $q \in V$ and $q = pv$, for some $v \in V$. Then $p = qv^{-1} \in VV^{-1} \subset U$, a contradiction.)

**5. Corollary.** Every topological group $(G, m, \tau)$ is a regular space.

**Proof.** From Definition 1(iii) and Lemma 3(b), we immediately get that $G$ is $T_1$. Let $p \in G$ and $U$ be any neighborhood of $p$. Then $L_{p^{-1}}(U)$ is a neighborhood of $e$. Therefore, by Lemma 4, there exists a neighborhood $V$ of $e$ such that $V^{-1} \subset L_{p^{-1}}(U)$. Then, letting $W = L_p(V)$, we get that $W$ is a neighborhood of $p$, with (see Lemma 3(b))

$$W^- = L_p(V^-) \subset U.$$

While the product of open subsets of a topological group is open, the product of closed sets may not be closed. For example, in $E^2$ let

$$A = \{(x, y) | x > 0 \text{ and } y \geq l/x\},$$

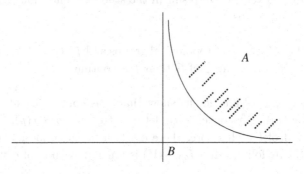

$$B = \{(0, y) | y \in E^{-1}\}.$$

Thinking of the points of $A$ and $B$ as vectors, it is easy to see that $A + B$ is the open right half-plane $\{(x, y)|y \in E^1 \text{ and } x > 0\}$. However, not all is lost.

**6. Lemma.** Let $G$ be a topological group, $B$ a closed subset of $G$, $C$ and $H$ compact subsets of $G$. Then

    (i) $CH$ is compact.

    (ii) $BC$ is closed.

    (iii) If $H$ is also a normal subgroup of $G$, then the natural quotient map $\lambda : G \to G/H$ is also a closed function.

**Proof.** Part (i) is immediate from Theorem 3.8 and Lemma 3.6(c).

Part (ii). Let $p \notin BC$. Then, for every $c \in C$, $e \notin p^{-1}Bc$ and $p^{-1}Bc$ is closed (since $p^{-1}Bc = R_c L_{p^{-1}}(B)$). Therefore, by Lemma 4, there exists a symmetric open neighborhood $V_c$ of $e$ such that

$$(*)V_c V_c \cap p^{-1}Bc = \emptyset.$$

Since $\{cV_c|c \in C\}$ is an open cover of $C$, let $\{c_1 V_{c_1}, \ldots, c_m V_{c_m}\}$ be a finite subcover, and let

$$V = \bigcup_{i=1}^{m} c_i V_{c_i}.$$

We show that $pV \cap BC = \emptyset$ (of course, this will complete the proof): Say $pv = bc$, for some $v \in V$, $b \in B$ and $c \in C$. Since $c \in \bigcup_{i=1}^{m} c_i V_{c_i}$, let $c = c_j v_j$, for some $j$ and $v_j \in V_{c_j}$. Then $pv = bc_j v_j$ or $vv_j^{-1} = p^{-1}bc_j \in p^{-i}Bc$, contradicting $(*)$.

Part (iii). All we need to show is that, for each closed $B \subset G$, $\lambda(B)$ is closed. But

$$\lambda^{-1}[\lambda(B)] = BH$$

is closed by (ii). Therefore, $\lambda(B)$ is closed, since $\lambda$ is a quotient map.

The following is a very useful result in the study of continuous homomorphisms between topological groups.

**7. Theorem.** Let $G$ and $H$ be topological groups and $f : G \to H$ a homomorphism. If $f$ is continuous at one point $p$ of $G$ then $f$ is continuous.

**Proof.** Pick any $q \in G$ and let us show that $f$ is continuous at $q$: Let $U$ be a neighborhood of $f(q)$. For convenience, let $a = f(q)$ and $b = f(p)$. Then $L_{ba^{-1}}(U)$ is a neighborhood of $b$. Therefore, there exists a neighborhood $V$ of $p$ such that $f(V) \subset L_{ba^{-1}}(U)$. It follows that $L_{qp^{-1}}(V)$ is a neighborhood of $q$ such that

$$f(L_{qp^{-1}}(V)) \subset U.$$

**8. Theorem.** Let $\{(G_\alpha, m_\alpha, \tau_\alpha)\}_{\alpha \in F}$ be a family of topological groups (finite, if you wish). Then $\Pi_{\alpha \in F} G_\alpha$ with the coordinatewise multiplication $m$ and the product topology $\Pi \tau_\alpha$, is a topological group.

**Proof.** It suffices to apply *old friends*—Lemma 2.7, Theorem 2.12—to the following commutative diagram

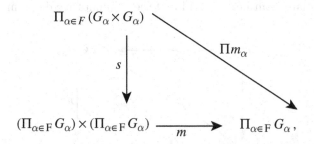

with $s((a_\alpha, b_\alpha)_\alpha) = ((a_\alpha)_\alpha, (b_\alpha)_\alpha)$, $m((a_\alpha)_\alpha, (b_\alpha)_\alpha) = (a_\alpha b_\alpha)_\alpha$, $\Pi m_\alpha((a_\alpha, b_\alpha)_\alpha) = (a_\alpha b_\alpha)_\alpha$. Clearly, the inversion map $i : \Pi_{\alpha \in F} G_\alpha \to \Pi_{\alpha \in F} G_\alpha$, with $i((a_\alpha)_\alpha) = (a_\alpha^{-1})_\alpha$, is also continuous, because of Lemma 2.7.

## 5.2 Topological Isomorphism Theorems

Let us recall that, for any homomorphism $\psi : G \to H$ between two groups, the *kernel of* $\psi$ is $Ker\, \psi = \{g \in G | \psi(g) = e\}$, where $e$ is the unit element of $H$. For convenience let $G \approx H$ denote that $G$ and $H$ are isomorphic (*i.e.* a bijective homeomorphism) and let $G \cong H$ denote that there exists a *topological isomorphism* (*i.e.*, an isomorphism and a homeomorphism) between $G$ and $H$.

**9. Lemma.** Let $G$ and $H$ be topological groups and $\psi : G \twoheadrightarrow H$ a quotient homomorphism. Then

$$G/Ker\, \psi \cong H.$$

**Proof.** Immediate from Lemma 2.14 and the commutative diagram with $j(g\, Ker\, \psi) = \psi(g)$, for every $g \in G$.

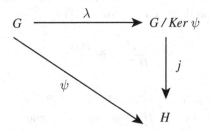

**10.  Theorem.** Let $\varphi : G \twoheadrightarrow G'$ be a quotient homomorphism and $H$ a closed normal subgroup of $G$ with $H \subset Ker\,\varphi$. Then

$$\frac{G}{Ker\,\varphi} \cong \frac{G/H}{Ker\,\varphi/H}.$$

**Proof.** Immediate from Lemma 2.14 and the commutative diagram

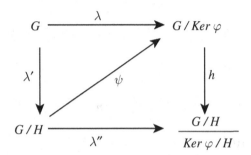

with $\psi(gH) = g\,Ker\,\varphi$, $\lambda''(gH) = gH(Ker\,\varphi/H) = g(Ker\,\varphi/H) = g\,Ker\,\psi$, $h(g\,Ker\,\varphi) = g(Ker\,\varphi/H) = g\,Ker\,\psi$; $\lambda$ and $\lambda'$ need no comment. It is straightforward that $h$ is an isomorphism. Since $\lambda$ and $\lambda'$ are quotient maps, it follows that $h$ and $h^{-1}$ are continuous, respectively—the diagrams may help.

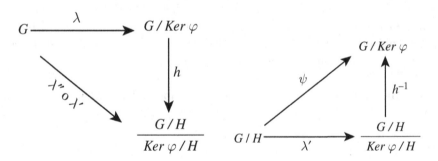

**11.  Theorem.** Let $H$ and $K$ be closed subgroups of a topological group $G$, with $K$ normal, $HK$ closed, and $H$ or $K$ compact. Then

$$\frac{H}{H \cap K} \cong \frac{HK}{K}.$$

**Proof.** Naturally, one first checks that $H \cap K$ and $HK$ are subgroups of $G$ and that $H \cap K$ is a normal subgroup of $H$. Next one constructs an appropriate diagram, that befits the situation at hand, and draws the appropriate conclusions:

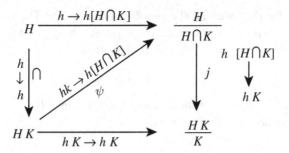

Note that $\psi$ is continuous whenever $H$ (or $K$) is compact. No further comments seem necessary. We leave the details to the reader.

## 5.3 Quotient Group Recognition

The usefulness of any mathematical structure is directly proportional to one's geometrical understanding of it. The applications of Chapter 3 offer overwhelming support for this statement. Fortunately, many a quotient group can also be *identified* with elementary groups. Let us treat two examples in order to illustrate the general techniques. The next chapter will have many more examples.

**12. Lemma.** The quotient group $E^1/Z$, with $Z$ being the additive group of the integers of $E^1$, is topologically isomorphic with the multiplicative group $S^1$ (*i.e.*, $S^1 = \{e^{i\theta} | 0 \leq \theta \leq 2\pi\}$ and $e^{i\theta} \times e^{i\beta} = e^{i(\theta+\beta)}$).

**Proof.** First, recall that $e^{i\theta} = e^{i(\theta+2\pi k)}$, for every integer $k$. Also recall that, for each $0 \leq t \leq 1$ and integer $k$, we get that the cosets

$$(k+t) + Z = t + Z.$$

Then the diagrams

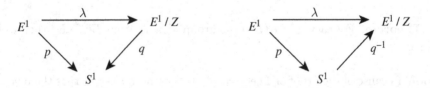

with $p(r) = e^{i2\pi r}$, $q(r + Z) = e^{i2\pi r}$, for each $r \in E^1$, suffice to complete the proof. (Clearly $q$ is an isomorphism; also $p$ is open and continuous—therefore $p$ is quotient. Now, the first diagram says that $q$ is continuous and the second says that $q^{-1}$ is continuous.)

**13. Lemma.** The quotient group $E^1 \times E^1/Z \times Z$ ($Z \times Z$ is called the group of Gaussian integers) is topologically isomorphic with the *torus* group $S^1 \times S^1$.

**Proof.** Essentially the same as the proof of Lemma 12. The only concern is to check that $S^1 \times S^1$ is (homeomorphic to) a torus: Simply define $\ell : I \to S^1$, by letting $\ell(t) = e^{i2\pi r}$, for every $t \in I$, and observe that

$$\ell \times \ell : I \times I \to S^1 \times S^1$$

produces exactly the same identifications of the boundary $\partial(I \times I)$ of $I \times I$ as the ones to construct the torus in Chapter 3.

## 5.4   Morphism Groups

By a *morphism* $f : X \twoheadrightarrow Y$ we mean a function with one or more of the following attributes: homeomorphism, isometry, toplogical isomorphism.

A *topological transformation group* (abbrev. ttg) is a pair $(G, X)$ such that $X$ is a topological space, $G$ is a topological group of morphisms $X \overset{f}{\twoheadrightarrow} X$ with respect to composition of functions, and the evaluation function $G \times X \to X$ (note that $G \subset X^X$) is continuous. Unless otherwise stated, *all ttgs will be assumed to have co topology.*

**14. Theorem.** For any space $X$, let $\mathcal{U}(X)$ be a group of morphisms $X \overset{f}{\twoheadrightarrow} X$ with respect to composition of functions. If $X$ is compact Hausdorff, then $\mathcal{U}(X)$ is a ttg.

**Proof.** Clearly, the evaluation map $\mathcal{E} : G \times X \to X$ is continuous, because of Corollary 4.13, and the group multiplication is continuous, because of Lemma 4.12. The inversion function $i : \mathcal{U}(X) \to \mathcal{U}(X)$ is also continuous. (Let $f^{-1} \in \langle K, V \rangle$. Then $f \in \langle X - V, X - K \rangle$ and $i(\langle X - V, X - K \rangle) \subset \langle K, V \rangle$.)

In the preceding result, we definitely used the compactness of $X$ to show that the inversion function is continuous. But do we really need compactness? If $X$ has some *good connectivity properties* then we don't need that $X$ be compact. Let us substantiate our vague statement with the Euclidean spaces.

**15. Theorem.** For each $n$, if $\mathcal{U}(E^n)$ is a group of morphisms $E^n \overset{f}{\twoheadrightarrow} E^n$ then $\mathcal{U}(E^n)$ is a ttg.

**Proof.** Because of the proof of Theorem 14, we only need to show that the inversion function is continuous. But, first we need to check that

$$\mathcal{K}_{co} = \{\langle K, V \rangle | K \text{ is compact connected, } V \text{ is open}\}$$

is also a subbase for the *co* topology: Clearly $\mathcal{K}_{co} \subset \mathcal{S}_{co}$. Now, let $f \in \langle C, V \rangle \in \mathcal{S}_{co}$. Since $f(C) \subset V$ and $C$ is compact, there exists finitely many balls $B(c_1, \delta_1), \ldots, B(c_n, \delta_n)$, with centers in $C$, which cover $C$ and such that

$$f(B(c_i, \delta_i)^-) \subset V.$$

Then, each $\langle B(c_i, \delta_i)^-, V \rangle \in \mathcal{K}_{co}$, and

$$f \in \bigcap_{i=1}^{n} \langle B(c_i, \delta_i)^-, V \rangle \subset \langle C, V \rangle.$$

This shows that

$$\mathcal{S}_{co} \subset \text{The topology generated by } \mathcal{K}_{co}.$$

It follows that $\mathcal{K}_{co}$ is a subbase for the *co* topology.

Now we show that the inversion function is continuous: let $f^{-1} \in \langle K, V \rangle \in \mathcal{K}_{co}$. Using the local compactness of $E^n$, the connectedness of balls in $E^n$, and Theorems 3.33 and 3.34, one can easily pick compact connected sets $K_*$ and $V_*$ such that

$$K \subset K_*^0, f^{-1}(K_*) \subset V_*^0 \subset V_* \subset V,$$

which implies that

$$f^{-1} \in \langle K_*, V_*^0 \rangle \subset \langle K, V \rangle.$$

Now let $V_* \subset B(p, \gamma)$, for some $p \in E^n$ and $\gamma > 0$ (cf. Theorem 3.9), and let $K' = B(p, \gamma)^- - V_*^0$. The following diagram should help in the ensuing argument.

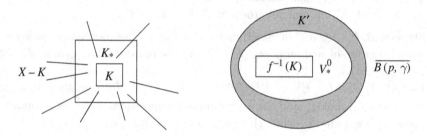

It is clear that $f \in N = \langle K', X - K \rangle \cap \langle f^{-1}(K), K_*^0 \rangle$; furthermore $i(N) \subset \langle K_*, V^-0_* \rangle \subset \langle K, V \rangle$: Suppose that there exists $g \in N$ such that $g^{-1}(K_*) \not\subset V_*^0$. Then, for every $q \in K_*$ such that $g^{-1}(q) \notin V_*^0$, we get that $g^{-1}(q) \notin B(p, \gamma)^-$, because $gg^{-1} = $ identity function. Therefore,

$$g^{-1}(K) \subset V_*^0 \cup [E^n - B(p, \gamma)^-],$$

$$g^{-1}(K) \cap V_*^0 \neq \emptyset \neq g^{-1}(K) \cap [E^n - B(p, \gamma)^-],$$

which implies that $g^{-1}(K)$ is not connected, a contradiction to Lemma 3.31.

## Chapter 5. Exercises

1. A topological space $X$ is said to be *homogeneous* if, for all $x, y \in X$, there exists a homeomorphism $h : X \twoheadrightarrow X$ such that $h(x) = y$. Show that

   (a) $I$ is not homogeneous (Hint: Let $x = 1$, $y = \frac{1}{2}$).

    (b) The subspace of $E^2$ which consists of the $x$-axis and $y$-axis is not homogeneous (Hint: Let $x = (0,0)$, $y = (1,0)$).

    (c) Every topological group is homogeneous.

    (d) $I$ is not a topological group.

2. Let $G$ be a topological group, $n \in \mathbf{N}$, and $U$ a neighborhood of $e \in G$. Show that there exists a symmetric neighborhood $W$ of $e$ such that $W^n \subset U$, where $W^1 = W$ and $W^n = W^{n-1}W$, for $n = 2, 3, \ldots$.

3. Show that an open subgroup of a topological group is also closed. If a topological group is connected what are its open subgroups?

4. Let $(G, m)$ be an algebraic group and $\tau$ a $T_1$-topology on $G$. Show that $(G, m, \tau)$ is a topological group iff the operation $\ell : G \times G \to G$, defined by $\ell(x,y) = xy^{-1}$, is continuous.

5. Let $(G, m, \tau)$ be a topological group and $H$ a subgroup of $G$ (not necessarily normal or closed). Show that $G/H$ is homogeneous (cf. ex. 1).

6. Let $X$ be any space and consider the operation $A : C(X, E^n) \times C(X, E^n) \to C(X, E^n)$, defined by $A(f,g)(x) = f(x) + g(x)$. Show that $(C(X, E^n), A, co.)$ and $(C(X, E^n), A, pc.)$ are topological groups.

7. Let $(G, m, \tau)$ be a topological group and $H$ a subgroup of $G$. Show that

    (a) $H^-$ is a (closed) subgroup of $G$.

    (b) If $H$ is a normal subgroup of $G$ then so is $H^-$.

8. Show that the logarithm function is a topological isomorphism between the additive group of real numbers $(E^1, +)$ and the multiplicative group of positive reals $(E^1_+, \times)$.

9. Let $A(E^1) = \{f : E^1 \to E^1 | f(x) = rx + s; r, s \in E^1, r \neq 0\}$. Show that, with the $co.$ topology and the operation of composition, $A(E^1)$ is a topological group. (This group is called the group of *affine transformations* of $E^1$.)

10. Let $G$ and $H$ be groups and $\lambda : G \twoheadrightarrow H$ be an epimorphism. Suppose that there exists a function $s : H \to G$ such that $\lambda(s(h)) = h$, for each $h \in H$ (the function $s$ is called a *cross-section* or a *selection* for $\lambda$). Let $\varphi : H \times \operatorname{Ker} \lambda \to G$ be defined by $\varphi(h, k) = s(h)k$. Show that

    (a) $\varphi$ is a bijection.

    (b) $\varphi$ is a homomorphism if either $s$ is a homomorphism or $ghg^{-1} = h$, for every $g \in G$ and $h \in \operatorname{Ker} \lambda$ (hence, $\varphi$ is an onto isomorphism, in either case).

    (c) If $G$ and $H$ are topological groups, then $\varphi$ is continuous. (Hint: Consider the diagram

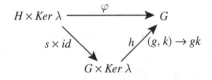

and check that $\varphi = h \circ (s \times id)$, with both these functions continuous.)

11. Let $(G, m, \tau)$ be a topological group and $h : G \twoheadrightarrow H$ a bijection. Show that $H$ can be given a topological group structure. (Hint: For all $a, b \in H$, let $ab = h(h^{-1}(a)h^{-1}(b))$.)

12. Prove that the Euclidean metric in $E^n$ satisfies the following.

   (a) $|\overline{tx} - \overline{ty}| = |t||\bar{x} - \bar{y}|$, for every $t \in E^1$.
   (b) $|(\bar{x} + \bar{z}) - (\bar{y} + \bar{z})| = |\bar{x} - \bar{y}|$ (*i.e.* the Euclidean metric is translation invariant).
   (c) Let $S$ be a subset of the plane $E^2$ such that $(0,0) \in S$; pick any $0 \le \theta \le 2\pi$. Show that $e^{i\theta}S$ is $S$ rotated through an angle $\theta$ about the origin $(0,0)$. (Hint: Think of the points $s$ of $S$ as being vectors starting at $(0,0)$ and ending at $s$. Recall that we can let $E^2 = \{re^{i\theta}|r, \theta \in E^1\}$ and that $r_1 e^{i\theta_1} r_2 e^{i\theta_2} = r_1 r_2 e^{i(\theta_1 + \theta_2)}$.)
   (d) Now show that (a) implies that the Euclidean metric in the plane is rotation invariant.
   (e) Show that (b) implies that

$$\int_{-\infty}^{\infty} f(x)dx = \int_{-\infty}^{\infty} f(x+r)dx = \int_{-\infty}^{\infty} f(-x)dx.$$

13. Let $(G, m)$ be an algebraic group and $\eta$ a family of subsets of $G$ such that

   (i) $\bigcap \eta = \{e\}$, where $e$ is the unit element of $G$,
   (ii) $M, N \in \eta$ implies $M \cap N \in \eta$,
   (iii) $M \subset N \subset G$, $M \in \eta$ implies $N \in \eta$,
   (iv) for each $N \in \eta$ there exists $M \in \eta$ such that $MM^{-1} \subset N$,
   (v) $N \in \eta$, $g \in G$ implies $gNg^{-1} \in \eta$.

   Prove that there exists a unique topology $\tau(\eta)$ for $G$ such that $(G, m, \tau(\eta))$ is a topological group and $\eta$ is the family of neighborhoods of $e$ with respect to $\tau(\eta)$. (Hint: Let

$$\tau(\eta) = \{U \subset G | x \in U \text{ implies that there exists } N_x \in \eta \text{ with } xN_x \subset U\}.$$

   To show that the multiplication is continuous: Say $xy \in xyU$, $U \in \eta$. Pick $V \in \eta$ such that $VV^{-1} \subset U$ and let $W = yVy^{-1} \cap V^{-1}$; note that (iv) implies that $M \subset N^{-1}$. Then show that $(xW)(yW) \subset xyU$.)

14. *Quaternions.* Consider $Q = E^2 \times E^2$ and the following operations on $Q$ (let $\bar{z}$ denote the conjugate of $z$; if $z = a + ib$ then $\bar{z} = a - ib$)

$$(x, y) + (w, z) = (x + w, y + z),$$

$$(x, y)(w, z) = (xw - \bar{y}z, yw + \bar{x}z),$$

   where we use the ordinary multiplication of complex numbers. For simplicity, we identify $c \in E^l$ with $(c, 0) \in E^2$. Show that

(i) $Q$ is a topological group with respect to addition, and $Q - \{0\}$ is a non-abelian topological group with respect to multiplication. (Show $(0,1)(i,0) \neq (C,0) \in E^2(0,1)$. The unit element is $(1,0)$. The multiplicative inverse of $(w,z) \in Q$ is $\left( \frac{\bar{w}}{w\bar{w}+z\bar{z}}, \frac{-z}{w\bar{w}+z\bar{z}} \right)$.

(ii) $E^1 - \{0\}$ is a closed, normal multiplicative subgroup of $Q - \{0\}$.

(iii) $E^2 - \{0\}$ is a closed, non-normal multiplicative subgroup of $Q - \{0\}$.

(iv) The *conjugation* function $\overline{(a,b)} = (\bar{a}, -b)$ in $Q$ is a continuous function.

(v) The *norm* function $\eta : Q - \{0\} \to E^1_+$, defined by $\eta(q) = q\bar{q}$, is an open epimorphism such that $\operatorname{Ker}\eta = S^3$. (This gives $S^3$ a non-abelian group structure.)

(vi) $Q - \{0\} \cong S^3 \times E^1_+$. (Hint: Use ex. 10.)

15. Let $H(]0,1[)$ be the set of all homeomorphisms of $]0,1[$ onto $]0,1[$, with the *co.* topology. Show that

(a) $H(]0,1[)$, with respect to composition of functions, is a ttg.

(b) For each $0 < x < 1$, $\gamma = \mathcal{E}|H(]0,1[) \times \{x\}$, where $\mathcal{E}$ is the evaluation function (cf. Section 4.4), is an open continuous function.

(c) For each $0 < x < 1$, $\gamma^{-1}(x) = H$, is a subgroup of $H(]0,1[)$. (This group is called the *isotropy* subgroup of $H(]0,1[)$ at $x$.)

(d) $H(]0,1[) \cong H_x \times ]0,1[$, for each $0 < x < 1$ (cf. ex. 10).

(e) $H([0,1[) \cong H(]0,1[)$.

16. Let $(G, m, \tau)$ be a topological group and $H(G)$ be the group (with respect to composition) of homeomorphisms of $G$ onto $G$ with the *co.* topology. Let $L : G \to H(G)$ be defined by $L(g) = L_g$. Show that:

(a) $L$ is continuous;

(b) $G$ is topologically isomorphic to a ttg (on $G$) which has the *co.* topology.

17. Let $(G, m, \tau)$ be a topological group and $U$ a neighborhood of $e \in G$. Show that the group generated by $U$ (*i.e.*, the smallest subgroup of $G$ which contains $U$) is $G$.

# Chapter 6

# Special Groups

Certain topological transformation groups have proved to be extremely important in the study of Quantum Mechanics, Relativity Theory and Crystallography. We will study a few of these. We try to follow a precise but very geometric approach.

## 6.1 Preliminaries

We will limit ourselves to $n \times n$-matrice $M$ over $E^1$. However, much of what will be done remains valid for $n \times n$-matrices over $E^2$.

Throughout, we will let

$$M = (m_{ij})_n = (m_{ij}) = \begin{pmatrix} m_{11}, \ldots, m_{1n} \\ \vdots \\ m_{n1}, \ldots, m_{nn} \end{pmatrix}$$

and we will think of $M$ as a function $M : E^n \to E^n$, defined by $M((x_1, \ldots, x_n)) = (\sum_j m_{1j}x_j, \ldots, \sum_j m_{nj}x_j)$, or in matrix-product form,

$$M \begin{pmatrix} x_1 \\ \vdots \\ x_n \end{pmatrix} = \left( \sum_j m_{1j}x_j, \ldots, \sum_j m_{nj}x_j \right).$$

From Theorem 2.3 and the basic properties of matrix multiplication, it follows easily that $M$ is a continuous linear (i.e., $M(t\bar{x} + s\bar{y}) = tM(\bar{x}) + sM(\bar{y})$, for all $s, t \in E^1$ and $\bar{x}, \bar{y} \in E^n$) function. Note that the identity $n \times n$-matrix $I_n$ (i.e., $I_n = (\delta_{ij})$ with $\delta_{ij} = 0$ whenever $i \neq j$ and $\delta_{ii} = 1$) is the identity isomorphism from $E^n$ to $E^n$. We will also think of $E^n$ as a vector space over $E^1$ with base $\{e_1, \ldots, e_n\}$, where $e_i$ is the $n$-tuple of $E^n$, whose only nonzero coordinate is the $i^{\text{th}}$-coordinate, which equals 1. Generally, we will not distinguish the point $\bar{a} = (a_1, \ldots, a_n) \in E^n$ from the line segment $\bar{a} = \{t\bar{a} | 0 \leq t \leq 1\}$ which joins the origin $(0, \ldots, 0)$ of $E^n$ to $(a_1, \ldots, a_n)$, inasmuch that this correspondence $(a_1, \ldots, a_n) \leftrightarrow \bar{a}$ is a bijection, and we will refer to $\bar{a}$ as an *n-vector*. Recall that the length of the $n$-vector $\bar{a}$ is denoted by $|\bar{a}| = \sqrt{\sum_i a_i^2}$. We will say that the two $n$-vectors $\bar{v}$ and $\bar{w}$ are orthogonal (abbrev. $\bar{v} \perp \bar{w}$) if the angle between them is 90°. A simple way of determining the angle between two $n$-vectors $\bar{a}$ and $\bar{b}$ is to use the law-of-cosines of trigonometry

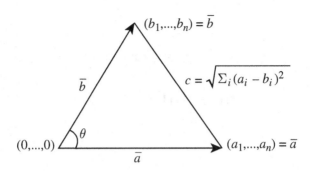

$$c^2 = |\bar{a}|^2 + |\bar{b}|^2 - 2|\bar{a}||\bar{b}|\cos\theta.$$

With $|\bar{a}|^2 = \sum_i a_i^2$ and $|\bar{b}|^2 = \sum_i b_i^2$, the preceding equation immediately yields that

$$\sum_i a_i b_i = |\bar{a}||\bar{b}|\cos\theta.$$

As customary, we call $\sum_i a_i b_i$ the *dot product* of the vectors $\bar{a}$ and $\bar{b}$ and let

$$\bar{a} \cdot \bar{b} = \sum_i a_i b_i.$$

We then get the elementary, but crucial, result:

**1. Proposition.** For any $n$-vectors $\bar{a}$ and $\bar{b}$, the following is true:

(i) $\bar{a} \cdot \bar{b} = |\bar{a}||\bar{b}|\cos\theta$, where $\theta$ is the angle between $\bar{a}$ and $\bar{b}$.
(ii) $\bar{a} \perp \bar{b}$ iff $\bar{a} \cdot \bar{b} = 0$.

For any $n \times n$-matrix $M = (a_{ij})_n$ and $a_{k\ell} \in M$, we let $M_{k\ell}$ be the $(n-1) \times (n-1)$-matrix obtained from $M$ by removing the $k^{\text{th}}$-row and $\ell^{\text{th}}$-column and we let

$$C_{k\ell} = (-1)^{k+\ell} M_{k\ell}$$

($C_{k\ell}$ is called the $(k,\ell)$-cofactor of $M$). Also, for any $n \times n$-matrix $M = (a_{ij})_n$, we let $M^T = (a_{ij}^T)_n$ such that $a_{ij}^T = a_{ji}$. ($M^T$ is called the transpose of $M$.) We will also denote the determinant of a matrix $M$ by $|M|$ and think of $|M|$ inductively defined by cofactors; that is,

$$\begin{vmatrix} a_{11} & a_{12} \\ a_{21} & a_{22} \end{vmatrix} = a_{11}a_{22} - a_{21}a_{12},$$

and

$$|(a_{ij})_n| = \sum_j a_{1j}|C_{1j}|, \text{ for } n \geq 3.$$

The following elementary results will be useful:

**2. Proposition.** The following is valid:

(i) For any $n \times n$-matrices $M$ and $N$,

$$|MN| = |M||N|, (MN)^T = N^T M^T, (MN)^{-1} = N^{-1} M^{-1}$$

whenever $M^{-1}$ and $N^{-1}$ exist,

(ii) An $n \times n$-matrix $M = (a_{ij})$ has an inverse $M^{-1}$ iff $|M| \neq 0$; indeed $M^{-1} = \frac{1}{|M|}(C_{ij}^T)_n$ (*i.e.*, $(C_{ij}^T)$ is the transpose of the matrix of cofactors of $M$).

For computational purposes, it is also very convenient to think of an $n \times n$-matrix $(m_{ij})_n$ as an element of $E^{n^2}$, by letting

$$(m_{ij})_n = (m_{11}, \ldots, m_{1n}, m_{21}, \ldots, n_{2n}, \ldots, m_{n1}, \ldots, m_{nn}).$$

Then, letting $\mathfrak{m}_n$ be the set of all $n \times n$-matrices over $E^1$, we get that $\mathfrak{m}_n$ has the Euclidean metric $d((a_{ij}), (b_{ij})) = [\sum_{ij}(a_{ij} - b_{ij})^2]^{1/2}$.

**3. Proposition.** In $\mathfrak{m}_n$, the Eucliean metric topology equals the *co* topology, (thinking of functions $M : E^n \to E^n$).

**Proof.** Let $(m_{ij}) \in \langle K, V \rangle$. Then, let $d((m_{ij})(K), E^n - V) =$ some $\delta > 0$. (Indeed, in any metric space $(X, \rho)$, $A$ compact, $B$ closed and $A \cap B = \emptyset$ imply that $\rho(A, B) > 0$: Suppose not. Then there exists $\{x_n\} \subset A$ and $\{y_n\} \in B$ such that $d(x_n, y_n) < \frac{1}{n}$, for all $n$. Pick a subsequence $\{x_{n_k}\}$ of $\{x_n\}$ and $x \in A$ such that $\lim_k x_{n_k} = x$. Then $\lim_n d(y_{n_k}, x) = 0$, which implies that $x \in B$, a contradiction.) Also, choose $D \geq 1$ such that $(m_{ij})(K) \subset B(\bar{0}, D)$. It follows that

$$B((m), \delta/D) \subset \langle K, V \rangle.$$

Conversely, consider any $B((a_{ij}), \varepsilon)$. It follows that $(a_{ij}) \in \langle \{(1, 0, \ldots, 0)\}, ]a_{11} - \varepsilon/n^2, a_{11} + \varepsilon/n^2[ \times E^1 \times \cdots \times E^1 \rangle = S_{11}$. Similarly, pick $S_{k\ell}, 1 \leq k, \ell \leq n$, with $(a_{ij}) \in S_{k\ell}$ (move 1 and the interval...). Then, $(a_{ij}) \in \bigcap_{k,\ell} S_{k\ell} \subset B((a_{ij}), \varepsilon)$.

**4. Proposition.** The determinant function det: $\mathfrak{m}_n \to E^1$, defined by $\det(M) = |M|$, is open and continuous.

**Proof.** Because of Proposition 3 and Lemma 2.5, simply think of $\mathfrak{m}_n$ as a subspace of $\Pi_{i=1}^{n^2} E^1$. Since addition and multiplication of real numbers are open and continuous functions, and $\det M$ is a sum of products of real numbers, it follows immediately that det is an open and continuous function.

## 6.2 Groups of Matrices

For $n = 1, 2, \ldots$, let

$$GL(n, E^1) = \{M \in \mathfrak{m}_n : |M| \neq 0\},$$
$$Spin(n, E^1) = \{M \in \mathfrak{m}_n : |M| = \pm 1\},$$
$$SL(n, E^1) = \{M \in \mathfrak{m}_n : |M| = 1\},$$
$$O_n = \{M \in \mathfrak{m}_n | M^T = M^{-1}\},$$
$$SO_n = \{M \in \mathfrak{m}_n | M^T = M^{-1} \text{ and } |M| = 1\}.$$

It is quite easy to see that these are algebraic groups (see Proposition 2). They are generally called the *General Linear* group, the *Spin* group, the *Special Linear* group, the *Orthogonal* group and the *Special Orthogonal* group, respectively.

**5. Theorem.** The following are valid:

    (i) The topological groups $GL(n, E^1)$, $Spin(n, E^1)$, $SL(n, E^1)$, $O_n$ and $SO_n$ are ttgs.

    (ii) $GL(n, E^1) \supset Spin(n, E^1) \supset O_n \supset SO_n$.

    (iii) $GL(n, E^1)/SL(n, E^1) \cong E^1 - \{0\}$.

**Proof.** Part (i) follows immediately from Theorem 5.15. To prove (ii), we only need show that $M^T = M^{-1}$ implies $|M| = \pm 1$: Note that

$$1 = |MM^{-1}| = |MM^T| = |M||M^T| = |M|^2,$$

since it is clear that $|M| = |M^T|$. To prove (iii) first check that $SL(n, E^1)$ is a closed normal subgroup of $GL(n, E^1)$. Then note that det: $GL(n, E^1) \to E^1 - \{0\}$ is a quotient homomorphism, because of Propositions 4 and 2(i). Then apply Lemma 5.9.

## 6.3 Groups of Isometries

The following elementary observation has profound consequences. Rarely, does so little mean so much: The *isometries of the $n$-sphere $S^n$ are in one-to-one correspondence with the isometries of $E^{n+1}$ which leave the origin of $E^{n+1}$ fixed* (Let $h : S^n \twoheadrightarrow S^n$ be an isometry (recall ex. 3.11). Let $\tilde{h} : E^{n+1} \twoheadrightarrow E^{n+1}$ be defined by $\tilde{h}(\bar{0}) = \bar{0}$ and $\tilde{h}(x) = |\bar{x}| h\left(\frac{\bar{x}}{|\bar{x}|}\right)$, for each $x \in E^{n+1} - \{\bar{0}\}$. Then $\tilde{h}$ is an isometry of $E^{n+1}$ which leaves the origin $\bar{0}$ fixed. Conversely, let $\tilde{g} : E^{n+1} \twoheadrightarrow E^{n+1}$ be an isometry with $\tilde{g}(\bar{0}) = \bar{0}$. Then, for each $\bar{x} \in S^n$, $|\tilde{g}(\bar{x})| = |\tilde{g}(\bar{x}) - \tilde{g}(\bar{0})| = |\bar{x} - \bar{0}| = 1$, which implies that $g = \tilde{g}|S^n$ is an isometry of $S^n$).

For convenience, for $n = 1, 2, \ldots$, let

$$GI_n = \{h : E^n \twoheadrightarrow E^n | h \text{ is an isometry}\},$$

$$SI_n = \{\tilde{h} \in GI^n | \tilde{h}(\bar{0}) = \bar{0}\} \cong \{g : S^{n-l} \twoheadrightarrow S^{n-1} | g \text{ is an isometry}\}.$$

It is clear that, with the *co* topology, $GI_n$ and $SI_n$ are ttgs.

**6. Theorem.** $O_n \cong SI_n$.

**Proof.** Once we show that $O_n$ and $SI_n$ are algebraically isomorphic, the remainder of the proof will become clear.

Let $h : S^{n-1} \twoheadrightarrow S^{n-1}$ be an isometry and $\tilde{h} : E^n \twoheadrightarrow E^n$ its corresponding isometry of $E^n$ which leaves $\bar{0}$ fixed. Let $f_i = h(e_i)$, for $i = 1, 2, \ldots, n$ (recall that the $e_i$ are the elements of the usual base for $E^n$). Using the law-of-cosines and the fact that the cosine function is one-to-one, for $0 \le \alpha \le \pi$, we immediately get that

(1) $f_i \perp f_j$, whenever $i \ne j$; each $|f_i| = 1$,

which implies that $\{f_1, \ldots, f_n\}$ is an orthonormal vector basis for $E^n$. Next, note that, for each $a \in E^1$, and $i = 1, 2, \ldots, n$,
$$\tilde{h}(ae_i) = a\tilde{h}(e_i) = af_i.$$
Next, observing that, for each $\bar{x} \in S^{n-1}$, $\bar{x} = (x_1, \ldots, x_n) = x_1 e_1 + \cdots + x_n e_n$, we show that $\tilde{h}(\bar{x}) = x_1 f_1 + \cdots + x_n f_n$: Suppose $\tilde{h}(\bar{x}) = z_1 f_1 + \cdots + z_n f_n$. Since $\tilde{h}$ is an isometry with $\tilde{h}(\bar{0}) = \bar{0}$, we get that

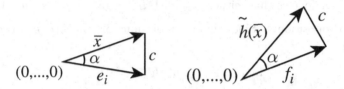

for $i = 1, 2, \ldots, n$. Therefore, we get the dot product equality
$$\bar{x} \cdot e_i = \tilde{h}(\bar{x}) \cdot f_i,$$
which yields $x_i = z_i$, for $i = 1, 2, \ldots, n$. This shows that $\tilde{h}$ is a linear map. Therefore, letting

(2) $f_i = \sum_j a_{ij} e_j = (a_{i1}, \ldots, a_{in})$,

for $i = 1, 2, \ldots, n$, and $M = (a_{ji})_n$, we immediately get that
$$\tilde{h}(\bar{x}) = M(\bar{x}),$$
for each $\bar{x} \in E^n$. We already know that the matrix of $\tilde{h}^{-1}$ is $M^{-1}$. To show that $M^{-1} = M^T$, first observe that

(3) $\sum_j a_{ij}^2 = 1, \sum_j a_{ij} a_{kj} = 0$, for $k \ne i$,

for $i = 1, 2, \ldots, n$, because of (1), (2) and Proposition 1. Therefore,
$$M^T(f_i) = (a_{ij}) \begin{pmatrix} a_{ij} \\ \vdots \\ a_{in} \end{pmatrix}$$
$$= \left( \sum_j a_{1j} a_{ij}, \ldots, \sum_j a_{ij} a_{ij}, \ldots, \sum_j a_{nj} a_{ij} \right)$$
$$= e_i, \text{ for } i = 1, 2, \ldots, n.$$

It follows that $\tilde{h}^{-1}(\bar{x}) = M^T(\bar{x})$, for each $\bar{x} \in E^n$. Since the inverse of a matrix is unique, we get that $M^{-1} = M^T$. Thinking of matrices as functions, we actually proved that $\tilde{h}$ amd $M$ are the same function on $E^n$.

Certainly the reader must be anxious to learn the real difference between the groups $O_n$ and $SO_n$ (we already know that

$$O_n = SO_n \cup \{M \in \mathfrak{m}_n | M^T = M^{-1} \text{ and } |M| = -1\}.$$

from Theorem 5(ii)). The proper gleaning of the proof of Theorem 6 will give us the answer.

**7. Theorem.** The groups $O_n$ and $SO_n$ are compact.

**Proof.** Because of Proposition 3, we only need to show that $O_n$ is a closed and bounded subset of $E^{n^2}$. The same applies to $SO_n$.

$O_n$ is closed in $E^{n^2}$: Say $\{M_k\}_k \subset O_n$ such that $\lim_k M_k = M$. Since each $|M_k| = \pm 1$, there exists a subsequence $\{M_{k_j}\}_j$ of $\{M_k\}$ such that either all $|M_{k_j}| = 1$ or all $|M_{k_j}| = -1$. It follows that, either $1 = \lim_j |M_{k_j}| = |M|$ or $-1 = \lim_j |M_{k_j}| = |M|$, because of Proposition 4. Since the inversion function in $GL(n, E^1)$ is continuous, it follows that

$$\lim_k (M_k)^T = \lim_k (M_k)^{-1} = M^{-1}.$$

But it is easily seen that $M^T = \lim_k (M_k)^T$ whenever $M = \lim_k (M_k)$. Therefore, $M^T = M^{-1}$ and $M = \pm 1$, which shows that $O_n$ is closed in $E^{n^2}$.

$O_n$ is bounded in $E^{n^2}$: Indeed from part (3) of the proof of Theorem 6, we immediately get that

$$\sum_{ij} a_{ij}^2 = \sum_i \sum_j a_{ij}^2 = n,$$

for each $(a_{ij})_n \in O_n$, which implies not only that $O_n$ is bounded but also that $O_n$ is contained in the sphere of $E^{n^2}$ with center $\bar{0}$ and radius $n$.

To simplify and clarify matters, let

$$AO_n = \{M \in O_n : |M| = -1\}$$

and $J_{k_1} \ldots k_m$, $1 \leq k_1 \leq \cdots \leq k_m \leq n$, be the $n \times n$-matrix which has the same entries as $I_n$ except that the $(k_1, k_1)$-, $\ldots$, $(k_m, k_m)$-entries equal $-1$ (note that this defines $J_k$, for $1 \leq k \leq n$). It is clear that

    (i) Each $J_{k_1 \ldots k_m} = J_{k_1} \cdots J_{k_m}$,
    (ii) $M \in SO_n$, $m$ odd implies $J_{k_1 \ldots k_n} M \in AO_n$,
    (iii) $M \in SO_n$, $m$ even implies $J_{k_1 \ldots k_m} M \in SO_n$,
    (iv) $J_{k_1 \ldots k_n}$ changes the vectors $e_{k_1}, \ldots, e_{k_m}$ to the vectors $-e_{k_1}, \ldots, -e_{k_m}$, leaving the remaining vectors fixed,

(v) For each $M \in AO_n$ and $1 \leq k \leq n$, there exists $M_* \in SO_n$ such that $J_k M_* = M$

(indeed, $M_*$ has the same entries of $M$ except that the $k$-column of $M_*$ has the negatives of the $k$-column of $M$).

The preceding trivial facts make the following result quite obvious.

**8. Theorem.** The following are valid:

(a) $O_n = SO_n \cup AO_n$, $SO_n \cap AO_n = \emptyset$ and, for $1 \leq m$ odd $\leq n$, $AO_n = \{J_{k_1 \ldots k_m} M | M \in SO_n\}$.

(b) For $n = 3$, $AO_3$ consists of those elements of $O_3$ which transform

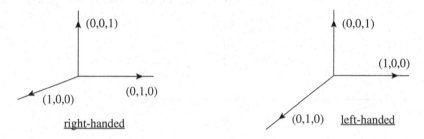

right-handed systems into left-handed systems and vice versa.

It is customary to say that a linear transformation $L : E^3 \to E^3$ *does not interchange past* (down) *with future* (up) if the *relative position of the z-axis with respect to the plane of the x-axis and y-axis* is not changed by $L$; that is, $L$ is of the form

$$L = R_x \circ R_y \circ J,$$

where $R_x (R_y)$ denotes a rotation about the x-axis (y-axis), and $J$ denotes a reversal of direction of the x-axis or of the y-axis or both, *i.e.*, $J$ is one of the matrices below

$$\begin{pmatrix} -1 & 0 & 0 \\ 0 & 1 & 0 \\ 0 & 0 & 1 \end{pmatrix}, \begin{pmatrix} 1 & 0 & 0 \\ 0 & -1 & 0 \\ 0 & 0 & 1 \end{pmatrix}, \begin{pmatrix} -1 & 0 & 0 \\ 0 & -1 & 0 \\ 0 & 0 & 1 \end{pmatrix}.$$

Similarly, we can speak of linear transformations which do not interchange *right with left* (*i.e.*, do not reverse the direction of the y-axis relative to the plane of the x-axis and z-axis); obviously, these concepts can be appropriately extended to higher dimensions.

## 6.4 Relativity and Lorentz Transformations

Our intention is to explain the why and how of Lorentz transformations in Relativity Theory.

Let us take two observers $O$ and $O'$ whose measurements are done with respect to the space-time coordinate systems $(x_1, x_2, x_3, X_4)$ and $(X_1, X_2, X_3, x_4)$, respectively, with the same origin $(0, 0, 0, 0) = \bar{0}$, where the first three coordinates are *space-coordinates* and the fourth is the *time-coordinate*. Furthermore, let us assume that $O$ and $O'$ are moving on a common straight line with constant speed *relative to each other*. Graphically, we may display the coordinate axis of $O$ and $O'$ as follows:

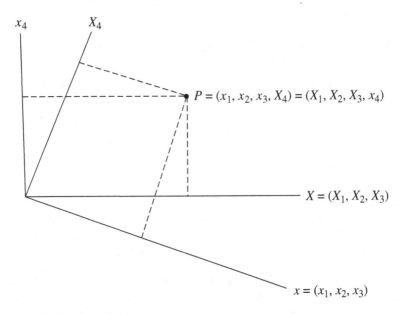

The apparent mislabeling of the axis reflects the fact that the space-time of $O$ is represented as seen by $O'$ and vice versa. Certainly the square of the distance of any point $P$ from the origin remains invariant for both observers; that is

$$x_1^2 + x_2^2 + x_3^2 + X_4^2 = X_1^2 + X_2^2 + X_3^2 + x_4^2$$

from which we get that

$$x_4^2 - \sum_{i=1}^{3} x_i^2 = X_4^2 - \sum_{i=1}^{3} X_i^2.$$

Our purpose is to identify the space-time linear transformations—*i.e.*, $4 \times 4$-matrices $A = (a_{ij})$—with respect to which the form

$$S(\bar{x}) = x_4^2 - \sum_{i=1}^{3} x_i^2$$

remains invariant, for any $\bar{x} = (x_1, x_2, x_3, x_4)$. First, let us observe that, letting

$$P = \begin{pmatrix} -1 & 0 & 0 & 0 \\ 0 & -1 & 0 & 0 \\ 0 & 0 & -1 & 0 \\ 0 & 0 & 0 & 1 \end{pmatrix} = (p_{ij}) = (p_{\alpha\beta}),$$

the invariance of $S$ can be described by

(1) $\sum_{i,j} p_{ij} X_i X_j = \sum_{\alpha,\beta} p_{\alpha\beta} x_\alpha x_\beta$.

Next, let us observe that, if $A((x_1, x_2, x_3, x_4)) = (X_1, X_2, X_3, X_4)$ then

(2) $X_m = \sum_n a_{mn} x_n$, for $m = 1, \ldots, 4$.

Therefore, substituting (2) in (1) yields

(3) $\sum_{i,j} p_{ij} (\sum_n a_{in} x_n)(\sum_k a_{jk} x_k) = \sum_{\alpha,\beta} p_{\alpha\beta} x_\alpha x_\beta$,

from which we get that

(4) $\sum_{\alpha,\beta} [\sum_{i,j} a_{i\alpha} p_{ij} a_{j\beta} - p_{\alpha\beta}] x_\alpha x_\beta = 0$,

because of the values that $p_{ij}$ takes. But (4) is true for all $\bar{x}$ iff

(5) $\sum_{i,j} a_{i\alpha} p_{ij} a_{j\beta} = p_{\alpha\beta}$, for each $\alpha$, $\beta$;

that is, iff

(6) $A^T P A = P$.

Then $\det(A^T P A) = \det P$ with $\det P = 1$. Therefore,

(7) $(\det A)^2 = 1$,

since $\det A^T = \det A$.

Also, letting $\alpha = \beta = 4$, we get that (5) becomes $a_{44}^2 - \sum_{i=1}^3 a_{i4}^2 = 1$, which implies that

(8) $a_{44}^2 \geq 1$.

We have therefore determined that

$$\mathcal{L} = \{(a_{ij})_{4\times4} = A | A \text{ satisfies (6), (7) and (8)}\}$$

is the set of all $4 \times 4$-matrices which leave the form $S$ invariant. It is obvious that if the linear transformations $A$, $B$ leave the form $S$ invariant, so do $AB$ and $A^{-1}$. Therefore, $\mathcal{L}$ is a subgroup of $Spin(n, E^1)$.

The group $\mathcal{L}$ is called the *full Lorentz group*. It splits into four subsets

$$\mathcal{L}_1 = \{A \in \mathcal{L} | \det A = 1, a_{44} \geq 1\}$$

$$\mathcal{L}_2 = \{A \in \mathcal{L} | \det A = 1, a_{44} \geq -1\}$$

$$\mathcal{L}_3 = \{A \in \mathcal{L} | \det A = -1, a_{44} \geq 1\}$$

$$\mathcal{L}_4 = \{A \in L | \det A = -1, a_{44} \geq -1\}$$

If $A \in \mathcal{L}_1 \cup \mathcal{L}_2$, $A$ is called a *proper Lorentz transformation*; otherwise, $A$ is an *improper Lorentz transformation*. If $a_{44} \geq 1$, $A$ does not interchange past with future; If $a_{44} \leq -1$, $A$ does interchange past with future. Finally, noting that

$$P(A^T P A) = P^2 = I$$

it is easy to compute the Lorentz matrix $L_v = (\ell_{ij})$ which satisfies the following natural physical constraints:

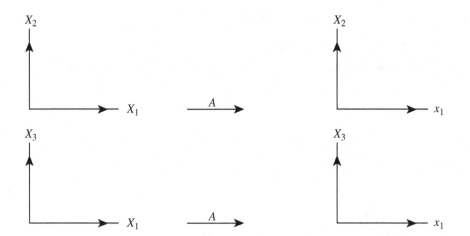

(i) It is required that the velocity of one coordinate system with respect to the other lies along a space-axis (say, the $X_1$-axis) and that the remaining orthogonal space-axis not be *interchanged* by $L_v$, thus precluding rotations about the $X_1$-axis; that is,

(ii) The relative velocity of observer $O'$ as seen by $O$ is a constant $v$. The velocity of light is taken to be *one unit*.

It follows that $X_1 = vX_4$, $X_2 = X_3 = 0$; furthermore,

$$\begin{pmatrix} x_1 \\ x_2 \\ x_3 \\ x_4 \end{pmatrix} = L_v \begin{pmatrix} 0 \\ 0 \\ 0 \\ x_4 \end{pmatrix} \quad \text{implies} \quad \begin{pmatrix} x_1 \\ x_2 \\ x_3 \\ x_4 \end{pmatrix} = \begin{pmatrix} \ell_{14}x_4 \\ \ell_{24}x_4 \\ \ell_{34}x_4 \\ \ell_{44}x_4 \end{pmatrix}$$

from which we get that

(a) $\ell_{14} = v\ell_{44} : \ell_{14}x_4 = X_1 = vX_4 = v\ell_{44}x_4$,
(b) $\ell_{24} = \ell_{34} = 0 : 0 = x_2 = \ell_{24}x_4$, $0 = X_3 = \ell_{34}x_4$.

Substitution into $PL_v^T = PL_v = I$ yields

$$\ell_{44}^2 - \ell_{14}^2 = 1 \text{ with } \ell_{14} = v\ell_{44}.$$

Therefore $\ell_{44}^2(1 - v^2) = 1$, or

(c) $\ell_{44} = \pm \dfrac{1}{\sqrt{1-v^2}} = \pm\gamma$

the minus sign signifying time reversal. Eliminating the time reversal situation, as unrealistic, by using (i) and (7), we can compute all other entries of $L_v$, finally getting

$$L_v = \gamma^{1/4} \begin{pmatrix} 1 & 0 & 0 & v\gamma \\ 0 & 1 & 0 & 0 \\ v & 0 & 1 & 0 \\ v & 0 & 0 & \gamma \end{pmatrix}$$

as the Lorentz transformation satisfying (i), (ii) and not reversing time. It is now clear that the most general Lorentz transformation has the form

$$A = \begin{cases} T \circ R \circ L_v \circ R', \\ P \circ R \circ L_y \circ R', \end{cases}$$

where $R$ and $R'$ are rotations and

$$T = \begin{pmatrix} -1 & 0 & 0 & 0 \\ 0 & -1 & 0 & 0 \\ 0 & 0 & -1 & 0 \\ 0 & 0 & 0 & 1 \end{pmatrix}$$

is the time-reversing matrix.

# Chapter 7

# Normality and Paracompactness

Some of the properties of metric spaces have proved so useful that they have been specially labeled and extensively studied. Among the dozens of significant properties of metric spaces, normality and paracompactness really stand out, and paracompactness outranks all others, by far. Let us discover these properties and some of their usefulness.

**1. Theorem.** Let $A$ and $B$ be closed disjoint subsets of a metric space $(X, \rho)$. Then there exist open disjoint subsets $U$ and $V$ of $(X, \rho)$ such that $A \subset U$ and $B \subset V$.

**Proof.** Define functions $f_A : X \to E^1$ and $f_B : X \to E^1$ by $f_A(x) = \rho(\{x\}, A)$ and $f_B(x) = \rho(\{x\}, B)$. Recall that $f_A$ and $f_B$ are continuous. (See ex. 1.20.) Next define $f : X \to E^1$ by $f(x) = \frac{f_A(x)}{f_A(x) + f_B(x)}$. Clearly $f$ is continuous (it is the ratio of continuous functions and the denominator is never zero. It is also clear that $f(A) = 0$ and that $f(B) = 1$. To complete the proof, let $U = f^{-1}(] - \frac{1}{2}, \frac{1}{2}[)$ and $V = f^{-1}(]\frac{1}{2}, 2[)$.

This result leads us to the definition of normal spaces.

## 7.1  Normal Spaces

**2. Definition.** A $T_1$-space $X$ is a *normal* space provided that for any disjoint closed subsets $A$ and $B$ of $X$ there exist disjoint open subsets $U$ and $V$ of $X$ with $A \subset U$ and $B \subset V$ (that is, disjoint closed subsets of $X$ can be separated by disjoint open subsets). Equivalently, for any closed subset $A$ of $X$ and open subset $U$ of $X$ with $A \subset U$, there exists open $V \subset X$ such that $A \subset V \subset V^- \subset U$.

**3. Corollary.** $X$ is metrizable implies that $X$ is normal. $X$ is normal implies that $X$ is regular. (The converses are false—see exs. 4 and 7(c,e).)

Recall that in the proof that the metric space is a normal space we constructed a continuous function $f : X \to E^1$ which mapped $A$ to 0 and $B$ to 1. Can we find such a function $f$ if $X$ is normal? This is indeed a deep question and the only known proof of it is truly ingenious. It was first discovered by Urysohn.

**4. Theorem (Urysohn's Lemma).** A $T_1$-space $X$ is normal if and only if for any disjoint closed subsets $A$ and $B$ of $X$ one can find a continuous map $f : X \to I$ such that $f(A) = 0$ and $f(B) = 1$.

**Proof.** The *if* part is contained in the proof of Theorem 1. The *only if* part: Let $D = U_n D_n$, where $D_0 = \{0, 1\}$, $D_1 = \{0, \frac{1}{2}, 1\}$, $D_2 = \{0, \frac{1}{4}, \frac{1}{2}, \frac{3}{4}, 1\}, \ldots$ $D$ is known as the set of dyadic rationals (note that these are obtained by dividing $[0, 1]$ into half, then the subintervals into half, and so on). It is easily seen that $D^- = I$ since any point of $I$ is less than $\frac{1}{2^n}$ away from some point of $D_n$.

To each $p \in D$ we will (inductively on $D_n$) associate an open subset $U_p$ of $X$ in such a way that

$$p < q \text{ implies } U_p^- \subset U_q.$$

$D_0$: Let $U_I = X$. Pick open $U_0 \subset X$ such that $A \subset U_0$ and $U_0^0 \cap B = \emptyset$.
$D_1$: Pick open $U_{\frac{1}{2}} \subset X$ such that

$$U_0^- \subset U_{\frac{1}{2}} \text{ and } U_{\frac{1}{2}}^- \cap B = \emptyset.$$

$D_2$: Pick open $U_{\frac{1}{4}}, U_{\frac{3}{4}} \subset X$ such that

$$U_0^- \subset U_{\frac{1}{4}} \subset U_{\frac{1}{4}}^- \subset U_{\frac{1}{2}} \subset U_{\frac{1}{2}}^- \subset U_{\frac{3}{4}} \text{ and } U_{\frac{3}{4}}^- \cap B = \emptyset.$$

The inductive procedure should now be clear. We may visualize the next step, on $D_3$, as follows:

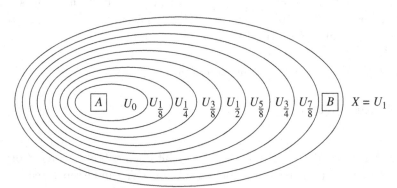

Now, we define $f : X \to [0, 1]$, by letting

$$f(x) = \inf\{p \in D | x \in U_p\}.$$

Clearly, $f$ is a well-defined function, and $f(A) = 0$ and $f(B) = 1$ (note that $f^{-1}(0) = U_0^- \neq A$; generally, one cannot expect that $f^{-1}(0) = A$; see exs. 16 and 17).

The easiest way to prove that $f$ is continuous is to use Theorem 1.14(vi). Note that

(i) $f^{-1}([0,t[) = \bigcup\{U_p | p < t\}$, for each $t \in I$: If $x \in U_P$ and $p < t$ then $f(x) \leq p < t$ which means that $x \in f^{-1}([0,t[)$; that is, $\bigcup\{U_p | p < t\} \subset f^{-1}([0,t[)$. Also, if $0 \leq f(x) < t$ then there exists $q \in D$ such that $f(x) < q < t$, which implies that $x \in U_q$; that is, $f^{-1}([0,t[) \subset \bigcup\{U_p | p < t\}$.

(ii) $f^{-1}([0,t]) = \subset \bigcap\{U_p^- | p < t\}$, for each $t \in I$: If $0 \leq f(x) \leq t$ and $p > t$ then $x \in U_p \subset U_p^-$; that is, $f^{-1}([0,t]) \subset \bigcap\{U_p^- | p > t\}$. If $x \in U_p^-$, for all $p > t$, then $x \in U_p$, for all $p > t$ (because, for $p > t$ with $p \in D$, there exists $q \in D$ such that $p > q > t$; hence, $x \in U_q \subset U_q^- \subset U_p$), which implies that $f(x) \leq t$; that is, $\bigcup\{U_p^- | p > t\} \subset f^{-1}([0,t])$.

(iii) $f^{-1}(]t,1]) = X - \bigcap\{U_p^- | p > t\}$: Immediate from (ii), since $f^{-1}(]t,1]) = X - f^{-1}([0,t])$.

Since the sets $[0,t[$ and $]t,1]$, for each $t \in I$, form a subbasis for the topology of $I$ and the inverse images, under $f$, of these sets are open (recall that any intersection of closed sets is closed), we then get that $f$ is continuous, which completes the proof.

Another very significant characterization of normality appears in exs. 24 and 25.

One may wonder if a similar result can be proved for regular spaces, with one of the sets replaced by a singleton. A look at the preceding proof shows that one cannot even get past the second induction step, for it may already require the separation of two non-degenerate closed sets. This suggests another definition: A $T_1$-space $X$ is *completely regular* or *Tychonoff* provided that, for each closed $A \subset X$ and $p \in X - A$, there exists a continuous function $f : X \to I$ such that $f(p) = 0$ and $f(A) = 1$. Clearly, $X$ is normal implies that $X$ is Tychonoff, which implies that $X$ is regular. It turns out that neither implication is reversible, but we shall not pursue this matter.

Let us now turn our attention to paracompactness. Be forewarned that the usefulness of this property of metrizable spaces is certainly matched by the difficulty in showing that metrizable spaces are indeed paracompact. All known proofs of this fact depend on the well-ordering Axiom (see 0.20).

## 5. Definition. In any space $X$,

(a) A cover $\mathcal{V}$ of $X$ (see Definition 3.4) is a *refinement* of a cover $\mathcal{U}$ of $X$ if each $V \in \mathcal{V}$ is contained in some $U \in \mathcal{U}$. $\mathcal{V}$ is an open (closed) refinement of $\mathcal{U}$ if $\mathcal{V}$ is an open (closed) cover of $X$ and a refinement of $\mathcal{U}$.

(b) A collection $\mathcal{U}$ of subsets of $X$ is *locally finite* if each $p \in X$ has a neighborhood which intersects only finitely many elements of $\mathcal{U}$.

(c) A collection $\mathcal{U}$ of subsets of $X$ is *discrete* if $\mathcal{U}$ is locally finite and pairwise disjoint (*i.e.*, for any distinct $U, V \in \mathcal{U}, U \cap V = \emptyset$).

(d) A collection $\mathcal{U}$ of subsets of $X$ is *$\sigma$-discrete* ($\sigma$-locally finite) if $\mathcal{U} = \bigcup_n \mathcal{U}_n$ and each $\mathcal{U}_n$ is discrete (locally finite).

Clearly finite collections of subsets of $X$ are locally finite and countable collections are $\sigma$-discrete. Pairwise disjoint collections may not be discrete. (The collection of intervals $]\frac{1}{n+1}, \frac{1}{n}[$, for $n = 1, 2, \ldots$, is not locally finite at the point $0$). One immediately gets that each compact Hausdorff space is paracompact. The converse is false (see Theorem 9 and ex. 10). Furthermore, if $\mathcal{U}$ is a locally finite collection of subsets of $X$ then $\mathcal{U}^- = \{U^-|U \in \mathcal{U}\}$ is also locally finite (note that if $N$ is an *open* neighborhood of a point $p \in X$ then $N \cap U \neq \emptyset$ iff $N \cap U^- \neq \emptyset$). Also, a *finite union of locally finite covers is clearly a locally finite cover*.

**6. Lemma.** The following are valid in any space $X$.

    (a) If $\mathcal{U} = \{U_\alpha|\alpha \in \Lambda\}$ is a locally finite collection of subsets of $X$ and, $V_\alpha \subset U_\alpha$, for each $\alpha \in \Lambda$, then $\mathcal{V} = \{V_\alpha|\alpha \in \Lambda\}$ is also locally finite.

    (b) If $\mathcal{A}$ is a locally finite collection of subsets of $X$ then $\bigcup\{A^-|A \in \mathcal{A}\} = (\bigcup \mathcal{A})^-$ (*i.e.*, the union of closures is the closure of the union).

**Proof.** Part (a) is obvious since $U \cap U_\alpha = \emptyset$ implies that $U \cap V_\alpha = \emptyset$ (that is, $U$ intersects no more elements of $\mathcal{V}$ than elements of $\mathcal{U}$).

    (b) We always have that $\bigcap\{A^-|A \in \mathcal{A}\} \subset (\bigcup \mathcal{A})^-$. So let $p \in (\bigcup \mathcal{A})^-$. Pick neighborhood $N$ of $p$ which intersects only finitely many $A \in \mathcal{A}$; say $A_1, \ldots, A_k$. Then $x \in (A_1 \cup \cdots \cup A_k) = A_1^- \cup \cdots \cup A_k^- \subset \bigcup\{A^-|A \in \mathcal{A}\}$ (recall ex. 1.21), which shows that $(\bigcup \mathcal{A})^- \subset \bigcup\{A^-|A \in \mathcal{A}\}$.

    Note that $]0, 1] = \bigcup_n [\frac{1}{n+1}, \frac{1}{n}] \neq (\bigcup_n]\frac{1}{n+1}, \frac{1}{n}[)^- = [0, 1]$, which shows the necessity of local finiteness in Lemma 6(b).

## 7.2  Paracompact Spaces

**7. Definition.** A Hausdorff space $X$ is a *paracompact* space provided that each open cover of $X$ has an open locally finite refinement.

**8. Theorem.** The following properties of a regular space are equivalent:

    (a) $X$ is paracompact.
    (b) Each open cover of $X$ has a $\sigma$-locally finite open refinement.
    (c) Each open cover of $X$ has a locally finite refinement.
    (d) Each open cover of $X$ has a closed locally finite refinement.

**Proof.** Clearly (a) implies (b), since a locally finite cover is automatically $\sigma$-locally finite.

    (b) implies (c): Let $\mathcal{U}$ be an open cover of $X$. By (b): there exists an open refinement $\mathcal{V}$ of $\mathcal{U}$ such that $\mathcal{V} = \bigcup_n \mathcal{V}_n$ and each $\mathcal{V}_n$ is locally finite. For each $n$, let $D_n = \bigcup \mathcal{V}_n$. Clearly, $\mathcal{D} = \{D_n|n \in \mathbf{N}\}$ covers $X$. Next, let $A_n = D_n - \bigcup_{i=1}^{n-1} D_i$. Then $\mathcal{A} = \{A_n|n \in \mathbf{N}\}$ is a locally finite refinement of $\mathcal{D}$. (Note that every $p \in X$ belongs to some $D_j$ and, consequently, there exists a smallest integer $k$ such that $p \in D_k$; that is, $p \in D_k - \bigcup_{i=1}^{k-1} D_i = A_k$. This shows that $\mathcal{A}$ is a refinement of $\mathcal{D}$.

Note that we have also proved that $D_k$ is a neighborhood of $p$ which can intersect only $A_1, \ldots, A_k$; that is, $\mathcal{A}$ is locally finite.) We now get that $\mathcal{W} = \bigcup_n \{V \cap A_n | V \in \mathcal{V}_n\}$ is a locally finite refinement of $\mathcal{V}$. (Note that each $p \in X$ belongs to some $A_n$ and therefore automatically to $\bigcup \mathcal{V}_n$, which forces $p \in V \cap A_n$, for some $V \in \mathcal{V}_n$. This shows that $\mathcal{W}$ is a refinement of $\mathcal{V}$. Also, if you pick a neighborhood $N$ of $p$ which intersects only finitely many elements of $\mathcal{V}_1 \cup \ldots \cup \mathcal{V}_n$, then $N \cap D_n$ intersects only finitely many elements of $\mathcal{W}$—it certainly cannot intersect any $V \cap A_k \in \mathcal{W}$, with $k > n$.)

(c) implies (d): (This is where we need regularity of $X$.) Let $\mathcal{U}$ be an open cover of $X$. For each $x \in X$, pick $U_x \in \mathcal{U}$ such that $x \in U_x$ and, by regularity of $X$, pick an open neighborhood $V_x$ of $x$ such that $x \in V_x^- \subset U_x$. Then $\mathcal{V} = \{V_x | x \in X\}$ is an open refinement of $U$ such that $\mathcal{V}^-$ is also a closed refinement of $\mathcal{U}$. By (c), let $\mathcal{W}$ be a locally finite refinement of $\mathcal{V}$. Then $\mathcal{W}^-$ is a closed locally finite refinement of $\mathcal{V}^-$ and hence of $\mathcal{U}$.

(d) implies (a): Let $\mathcal{U}$ be an open cover of $X$ and $\mathcal{V}$ a closed locally finite refinement of $\mathcal{U}$. For each $x \in X$, let $O_x$ be a neighborhood of $x$ which intersects only finitely many $V \in \mathcal{V}$. Then, let $\mathcal{A}$ be a closed locally finite refinement of $\{0_x | x \in X\}$. For each $V \in \mathcal{V}$, let $V' = X - \bigcup\{A \in \mathcal{A} | A \cap V = \emptyset\}$. Then each $V'$ is an open set, by Lemma 6, and $V' \supset V$. Also $A \in \mathcal{A}$ intersects $V'$ iff $A$ intersects $V$; that is, $\mathcal{V}' = \{V' | V \in \mathcal{V}\}$ is locally finite. Consequently $\mathcal{V}'$ is an open locally finite cover of $X$ (not necessarily a refinement of $\mathcal{U}$!).

Now, for each $V \in \mathcal{V}$, pick $U_v \in \mathcal{U}$ such that $V \subset U_v$. Then $\mathcal{W} = \{U_v \cap V' | V \in \mathcal{V}\}$ is an open locally finite refinement of $\mathcal{U}$ (one can only question if $\bigcup \mathcal{W} = X$, but each $x \in$ some $V \subset V'$ and $V \subset U_v$ which implies that $x \in$ some $U_v \cap V' \in \mathcal{W}$).

Other characterizations of paracompactness are described in exs. 22 and 23.

## 9. Theorem. Every metric space $(X, d)$ is paracompact.

**Proof.** Let $\mathcal{U}$ be an open cover of $(X, d)$. For each $U \in \mathcal{U}$ and $n \in \mathbf{N}$ let $U_n = \{x \in U | d(\{x\}, X - U) \geq 2^{-n}\}$ (note that $\bigcup_n U_n = U$, because of Proposition 1.11(v); also, for *small* sets $U$, $U_n$ may be empty for *small* $n$). By the triangle inequality, we get that $d(U_n, X - U_{n+1}) \geq 2^{-n} - 2^{-n-1} = 2^{-n-1}$ (note that, for any $q \in U_n$, $p \in X - U_{n+1}$ and $z \in X - U$, we get that $d(q, p) \geq d(q, z) - d(z, p)$ with $d(q, z) \geq 2^{-n}$ and $d(z, p) < 2^{-n-1}$).

Let $\preceq$ be a well-order on $\mathcal{U}$ (see 0.20). For each $U \in \mathcal{U}$ and $n \in \mathbf{N}$, let $U'_n = U_n - \bigcup\{V_{n+1} | V \in \mathcal{U} \text{ and } V \preceq U\}$. Note that, for each $U, V \in \mathcal{U}$ and each $n$, either $U'_n \subset X - V_{n+1}$ or $V'_n \subset X - U_{n+1}$ (for example, $U \preceq V$ implies $V'_n \subset X - U_{n+1}$). In either case, $d(U'_n, V'_n) \geq 2^{-n-1}$ (for example, $U \preceq V$ implies $d(U'_n, V'_n) \geq d(U_n, X - U_{n+1}) \geq 2^{-n-1}$). Finally, for each $U \in \mathcal{U}$ and for each $n \in \mathbf{N}$, let $U^*_n = \{x | d(\{x\}, U'_n) < 2^{-n-3}\}$; it follows that $d(U^*_n, V^*_n) \geq 2^{-n-2}$ (note that, for any $q \in U^*_n$, $p \in V^*_n$, $z \in U'_n$ and $w \in V'_n$, we get that $d(q, p) \geq d(z, w) - d(z, q) - d(p, w) \geq 2^{-n-1} - 2^{-n-3} - 2^{-n-3} = 2^{-n-2}$, by the triangle inequality).

Then each $\mathcal{U}_n^* = \{U_n^*|U \in \mathcal{U}\}$ is an open discrete collection (indeed any ball of radius $< 2^{-n-2}$ can intersect at most one element of $\mathcal{U}_n^*$). Clearly, each $U_n^* \subset U$ (indeed $U_n^* \subset \{x \in U|d(\{x\}, X - U) > 2^{-n} - 2^{-n-3} > 2^{-n-3}\}$). Also, for each $x \in X$, if we let $U$ be the first (with respect to the well-order $\preceq$) element of $\mathcal{U}$ that contains $x$, then $x \in$ some $U_n^*$ (indeed, $x \in$ some $U_n$ and therefore $x \in U_n' \subset U_n^*$). Consequently, $\mathcal{U}^* = \bigcup_n \mathcal{U}_n^*$ is an open $\sigma$-discrete (hence, $\sigma$-locally finite) refinement of $\mathcal{U}$. By Theorem 8, $X$ is paracompact.

The preceding proof contains the very hard half of a characterization of metrizable spaces (see ex. 21).

**10. Theorem.** Every paracompact space $X$ is a normal space.

**Proof.** First, we show that $X$ is regular: Let $A$ be a closed subset of $X$ and $p \in X - A$. Since $X$ is Hausdorff, for each $a \in A$, pick an open neighborhood $N_a$ of $a$ such that $p \notin N_a^-$. Then, $\eta = \{N_a|a \in A\} \cup \{X - A\}$ is an open cover of $X$. So let $\mathcal{V}$ be an open locally finite refinement of $\eta$ and $\mathcal{A} = \{V \in \mathcal{V}|V \subset \text{ some } N_a \in \eta\}$. Clearly $A \subset \bigcup \mathcal{A}$ (i.e., $\mathcal{A}$ covers $A$), $A$ is open (union of open sets!) and $p \notin A^-$, by Lemma 6(b). This shows that $X$ is regular.

Finally, we show that $X$ is normal: Let $A$ be a closed subset of $X$ and $U$ an open subset of $X$ such that $A \subset U$. For each $a \in A$ pick an open neighborhood $N_a$ of $a$ such that $N_a^- \subset U$ (regularity!) and mimick the preceding scheme to complete the proof.

**11. Lemma.** Every open cover $\mathcal{U} = \{U_\alpha\}_{\alpha \in \Lambda}$ of a paracompact space $X$ has an open refinement $\mathcal{V} = \{V_\alpha\}_{\alpha \in \Lambda}$ such that each $\emptyset \neq V_\alpha^- \subset U_\alpha$.

**Proof.** Since $X$ is regular, let $\mathcal{W}$ be an open refinement of $\mathcal{U}$ such that $\mathcal{W}^- = \{W^-|W \in \mathcal{W}\}$ also refines $\mathcal{U}$. Then let $\mathcal{O}$ be an open locally finite refinement of $\mathcal{W}$. For each $\alpha \in \Lambda$, let $\mathcal{O}_\alpha = \{V \in \mathcal{O}|V \subset U_\alpha\}$.

Finally, we define the $V_\alpha$: If $\mathcal{O}_\alpha \neq \emptyset$, let $V_\alpha = \bigcup \mathcal{O}_\alpha$. If $\mathcal{O}_\alpha = \emptyset$, let $V_\alpha$ be any open subset of $X$ such that $\emptyset \neq V_\alpha \subset v_\alpha^- \subset U_\alpha$. From Lemma 6(b), one easily sees that $\mathcal{V} = \{V_\alpha\}_{\alpha \in \Lambda}$ satisfies all requirements.

Lemma 11 remains valid for every open *locally finite* cover $\mathcal{U} = \{U_\alpha\}_{\alpha \in \Lambda}$ of a *normal* space, but the proof of this fact is quite complicated (see ex. 11).

**12. Definition.** A *partition of unity* on a space $X$ is a collection $\mathcal{Q}$ of continuous functions from $X$ to $E_+^1$ (the non-negative reals) such that $\sum_{p \in \mathcal{Q}} p(x) = 1$, for each $x \in X$ (here, we automatically assume that, for each $x \in X$, $p(x) \neq 0$ for only finitely many $p \in \mathcal{Q}$). $\mathcal{Q}$ is called *locally finite* if each $x \in X$ has a neighborhood $N_x$ such that $p(N_x) = 0$, for all but finitely many $p \in \mathcal{Q}$. $\mathcal{Q}$ is *subordinated* to a cover $\mathcal{U}$ of $X$ if for each $p \in \mathcal{Q}$ there exists $U \in \mathcal{U}$ such that $p(X - U) = 0$ (i.e., $p$ *vanishes outside* $U$).

For example, let $(X, d)$ be a metric space and $\mathcal{U} = \{U_\alpha\}_{\alpha \in \Lambda}$ be an open locally finite cover of $X$. For each $\alpha \in \Lambda$, let $f_\alpha : X \to E^1_+$ be defined by $f_\alpha(x) = d(x, X - U_\alpha)$. From ex. 1.20 we get that each $f_\alpha$ is continuous. Clearly, $\{f_\alpha\}_{\alpha \in \Lambda}$ is subordinated to $\mathcal{U}$. Finally let $p_\beta(x) = \frac{f_\beta(x)}{\sum_{\alpha \in \Lambda} f_\alpha(x)}$, for each $\beta \in \Lambda$ and $x \in X$. It is easily seen that each $p_\beta$ is continuous (note that, locally, $\sum_{\alpha \in \Lambda} f_\alpha(x)$ is a finite sum), $\sum_{\beta \in \Lambda} p_\beta(x) - \sum_{\beta \in \Lambda} \frac{f_\beta(x)}{\sum_{\alpha \in \Lambda} f_\alpha(x)} = \frac{\sum_{\beta \in \Lambda} f_\beta(x)}{\sum_{\alpha \in \Lambda} f_\alpha(x)} = 1$ and $\{p_\alpha\}$ is subordinated to $\{U_\alpha\}$. Therefore $\{p_\alpha\}$ is a partition of unity subordinated to $\{U_\alpha\}$.

This very important result for metric spaces provides another characterization of paracompactness.

**13. Theorem.** For a $T_1$-space $X$, the following are equivalent:

    (a) $X$ is paracompact.

    (b) Every open cover of $X$ has a locally finite partition of unity subordinated to it.

    (c) Every open cover of $X$ has a partition of unity subordinated to it.

**Proof.** (a) implies (b): let $\mathcal{W}$ be an open cover of $X$ and $\mathcal{U} = \{U_\alpha\}_{\alpha \in \Lambda}$ be an open locally finite refinement of $\mathcal{W}$. By Lemma 11, let $\{V_\alpha\}_{\alpha \in \Lambda}$ be a cover of $X$ such that each $\emptyset \neq V_\alpha^- \subset U_\alpha$. For each $\alpha \in \Lambda$, pick a continuous function $f_\alpha : X \to I$ such that $f_\alpha(V_\alpha^-) = 1$ and $f_\alpha(X - U_\alpha) = 0$. Now, let $P_\alpha : X \to I$ be defined by $p_\alpha(x) = \frac{f_\alpha(x)}{\sum_{\beta \in \Lambda} f_\beta(x)}$, for each $x \in X$. (Clearly, $p_\alpha(x)$ is continuous because, locally, $\sum_\beta f_\beta(x)$ is a finite sum.) It follows that $\mathcal{Q} = \{p_\alpha\}_{\alpha \in \Lambda}$ is a locally finite partition of unity subordinated to $\mathcal{U}$; therefore $\mathcal{Q}$ is also subordinated to $\mathcal{W}$.

    (b) implies (c): This is obvious.

    (c) implies (a): We are going to use Theorem 8; hence, we must first check that $X$ is regular. (Let $U$ be an open subset of $X$ and let $u \in U$. Then, $\{U, X - \{u\}\}$ is an open cover of $X$; hence, there exists a partition of unity $\{p_1, p_2\}$ subordinated to it; say $p_1(X - U) = 0$ and $p_2(u) = 0$. Then, $p_1(u) = 1$, because $p_1(u) + p_2(u) = 1$. Then $V = p_1^{-1}(]\frac{1}{2}, 1[)$ is an open neighborhood of $u$ whose closure is contained in $U$; indeed, $V^- \subset p_1^{-1}([\frac{1}{2}, 1]) \subset U$.)

Let $\mathcal{U}$ be an open cover of $X$ and $\mathcal{Q}$ a partition of unity subordinated to $\mathcal{U}$. For each $i \in \mathbf{N}$, let $\mathcal{V}_i$ be the collection of all sets of the form $S(p, i) = \{x \in X | p(x) > \frac{1}{i}\}$, with $p \in \mathcal{Q}$. Let $\mathcal{V} = \bigcup_{i=1}^\infty \mathcal{V}_i$. Clearly $\mathcal{V}$ is an open refinement of $\mathcal{U}$. (Certainly, each $S(p, i) = p^{-1}(]\frac{1}{i}, 1])$ is open. Also, for each $x \in X$, there exists some $p \in \mathcal{Q}$ such that $p(x) \neq 0$ and this $p$ vanishes outside of some $U \in \mathcal{U}$. Then $x \in S(p, j) \subset V \in \mathcal{U}$.)

Also, each $\mathcal{V}_i$ is locally finite: Let $x_0 \in X$. Pick $p_1, \ldots, p_n \in \mathcal{Q}$ such that $p_1(x_0) + \cdots + p_n(x_0) > 1 - \frac{1}{2i}$ (recall that $\sum_{p \in \mathcal{Q}} p(x_0) = 1$ and $p(x_0) \neq 0$ for only finitely many $p \in \mathcal{Q}$) and then pick a neighborhood $N$ of $x_0$ such that $p_1(x) + \cdots + p_n(x) > 1 - \frac{1}{i}$, for all $x \in N$ (note that $p_1 + \cdots + p_n$ is a continuous function). It follows that $N$ intersects only the elements $S(p_1, i), \ldots, S(p_n, i)$ of $\mathcal{V}_i$ (suppose that $N$ intersects some $S(p, i)$ with $p \neq p_1, \ldots, p_n$; then $(p_1(x) + \cdots + p_n(x)) + p(x) > (1 - \frac{1}{i}) + \frac{1}{i} = 1$, a contradiction) and, therefore, that $\mathcal{V}_i$ is locally finite.

Consequently, $\mathcal{V}$ is an open $\sigma$-locally finite refinement of $\mathcal{U}$. This implies that $X$ is paracompact, by Theorem 8(b).

Partitions of unity are extremely useful in many areas of mathematics. They are crucial in the *theory of continuous* (differentiable) *extensions* of *continuous* (differentiable) *functions*. (A bit of this theory, but a very important one, is Tietze's Extension Theorem, which appears in exs. 24 and 25.) They are also very useful in embeddings but, unfortunately, these and other applications of partitions of unity are extremely technical. We content ourselves with an interesting, but weak, application.

**14. Definition.** A Hausdorff space $X$ is called an $m$-manifold if each $x \in X$ has an open neighborhood that is homemorphic to the $m$-Euclidean space $E^m$ (or, equivalently, to the open $m$-ball $B(\bar{0}, 1)$).

The projective plane, the Klein bottle and the 2-sphere $S^2$ are examples of 2-manifolds (of course, $E^2$ is another example). The Möbius band is not a 2-manifold, because it has *boundary* or *edge* points. It is known as a 2-*manifold with boundary*.

**15. Theorem.** If $X$ is a compact $m$-manifold then $X$ can be embedded in $E^n$, for some $n$.

**Proof.** Let $\{U_1, \ldots, U_n\}$ be a finite open cover of $X$ such that each $U_i$ is homeomorphic to $E^m$; say $g_i : U_i \twoheadrightarrow E^m$ is a homeomorphism. Since $X$ is paracompact, by Theorem 13, let $\{p_1, \ldots, p_n\}$ be a partition of unity subordinated to $\{U_1, \ldots, U_n\}$; without loss of generality, let us assume that each $p_i(X - U_i) = \bar{0} = (0, 0, \ldots, 0) \in E^m$. For each $i$, let $h_i : X \to E^m$ be defined by

$$h_i(x) = \begin{cases} p_i(x)g_i(x) & \text{for } x \in U_i, \\ \bar{0}, & \text{for } x \in X - U_i. \end{cases}$$

(Note that $h_i(x)$ is a product of a real number $p_i(x)$ and an $m$-tuple $g_i(x) \in E^m$). Clearly each $h_i$ is a continuous function.

Finally, let $\psi : X \to (\Pi_{i=1}^n E^1) \times (\Pi_{i=1}^n E^m)$ be defined by

$$\psi(x) = (p_1(x), \ldots, p_n(x), h_1(x), \ldots, h_n(x)).$$

Clearly $\psi$ is continuous (see Theorem 2.3). Next we show that $\psi$ is one-to-one: Suppose $\psi(x) = \psi(y)$. Then $p_i(x) = p_i(y)$ and $h_i(x) = h_i(y)$, for $i = 1, \ldots, n$. Since $\sum p_i(x) = 1$, there exist some $j$ such that $p_j(y) = p_j(x) > 0$, which implies that $x, y \in U_j$. Then

$$p_j(x)g_j(x) = h_j(x) = h_j(y) = p_j(y)g_j(y)$$

which implies that $g_j(x) = g_j(y)$. Since $g_j$ is a homeomorphism, we get that $x = y$; that is, $\psi$ is one-to-one.

Clearly, $\psi$ is a quotient function (see Theorem 3.7). Therefore one easily sees that $\psi$ is a homeomorphism (see ex. 2.13).

This result is indeed a weak one. It is known that any compact $m$-manifold can be embedded in $E^{2m+1}$. It is easily seen that, in Theorem 15, $n > 2m + 1$. For example, the 2-sphere $S^2$ is the union of two open sets $U_1$ and $U_2$ which are homomorphic to $E^2$ (for example, let $U_1 = S^2 - \{$south pole$\}$ and $U_2 = S^2 - \{$north pole$\}$). Then, by Theorem 15, $S^2$ is embedded in $E^6$. Of course, $S^2$ is naturally embedded in $E^3$.

Also note that the proof of Theorem 15, with obvious changes, also proves that *every compact Hausdorff locally Euclidean space $X$ (i.e.*, each point of $X$ has an open neighborhood which is homeomorphic to some Euclidean space—not necessarily the same one for all points) *is* (homeomorphic to) *a subspace of some Euclidean space.*

## Chapter 7. Exercises

1. Show that a $T_1$-space is normal iff each finite open cover $\mathcal{U}$ of $X$ has a finite open refinement $\mathcal{V}$ such that $\mathcal{V}^- = \{V^-|V \in \mathcal{V}\}$ also refines $\mathcal{U}$. (Hint: Use induction.)
2. Show that a closed subspace of a normal space is normal. (False for open subspaces—see ex. 8(c).)
3. Show that a closed subspace of a paracompact space is paracompact. (False for open subspaces—see ex. 8(c).)
4. Let $\hat{X}$ be the space of ex. 1.27. We already know that $\hat{X}$ is not metrizable. Show that $\hat{X}$ is normal. (Hint: Let $A$ and $B$ be disjoint closed subsets of $\hat{X}$. There are two cases to consider. Case 1. $p \notin A \cup B$: Then $A$ and $B$ are also open sets. Case 2. $p \in A$ (hence $p \notin B$). Then $B$ is finite and open and $\hat{X} - B$ is also open ....)
5. Let $X$ be a regular Lindelöf space. Show that $X$ is paracompact. (Hint: Theorem 8(b).)
6. Show that a paracompact separable space is Lindelöf. (Hint: Let $D$ be a countable dense subset of $X$. Let $\mathcal{U}$ be any open cover of $X$ and $\mathcal{V}$ an open locally finite refinement of $\mathcal{U}$. Show that $\mathcal{V}$ is countable: In how many $V \in \mathcal{V}$ can an element of $D$ be? Then find a countable subcover of $\mathcal{U}$: For each $V \in \mathcal{V}$ pick $U_v \in \mathcal{U}$ such that $V \subset U_v$.)
7. Let $(E^1, \tau_s)$ be the Sorgenfrey line (see ex. 1.3). Show that $(E^1, \tau_s)$ is paracompact. (Hint: See ex. 1.28 and 5.)
8. Again, let $\hat{X}$ be the space of ex. 1.27.

   (a) Show that $\hat{X}$ is compact Hausdorff; hence, paracompact and normal.
   (b) $\hat{X} \times \hat{X}$ is compact Hausdorff; hence, paracomact and normal.
   (c) The subspace $Y = \hat{X} \times \hat{X} - \{(p,p)\}$ of $\hat{X} \times \hat{X}$ is not normal. (Hint: let $A = (\hat{X} - \{p\}) \times \{p\}$ and $B = \{p\} \times (\hat{X} - \{p\})$. Note that a neighborhood $U$ of $B$ consists of all points of $Y - A$, except for finitely many points of each *horizontal line* $\hat{X} \times \{u\}$ with $u \in \hat{X} - \{p\}$.

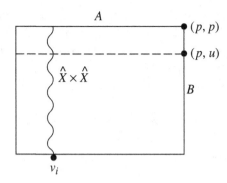

Pick a countably infinite subset $S = \{v_1, \ldots, v_n, \ldots\}$ of $\hat{X} - \{p\}$ and let $V_i = \{v_i\} \times \hat{X} - \{\text{finitely many points of } \hat{X} \times \hat{X}\}$, for $i \in \mathbf{N}$. Then $V = \bigcup_n V_n = S \times \hat{X} - \{\text{countably many points of } \hat{X} \times \hat{X}\}$. Now, show that $U \cap V \neq \emptyset$. Therefore, each neighborhood of $A$ intersects each neighborhood of $B$.

(d) $Y$ is open in $\hat{X} \times \hat{X}$. (Therefore open subsets of normal or paracompact spaces are not always normal or paracompact!)

(e) $Y$ is regular (see ex. 3.22).

9. (Variations on Urysohn's Lemma.) Show that Urysohn's Lemma remains valid if

   (a) $I$ is replaced by any closed interval $[a, b]$. (Hint: Use the homeomorphism $h : I \to [a, b]$, defined by $h(t) = ta + (1 - t)b$.)

   (b) $I$ is replaced by any open interval $]a, b[$, 0 and 1 are replaced by some points $p, q \in ]a, b[$, respectively.

10. Show that the real line $E^1$, with the usual topology is not compact (try the cover $\{]-n, n[ \,|\, n \in \mathbf{N}\}$), but it is paracompact.

11. Prove that, for a $T_1$-space $X$, the following are equivalent:

   (a) $X$ is normal.

   (b) Every locally finite open cover $\mathcal{U} = \{U_\alpha\}_{\alpha \in \Lambda}$ of $X$ has an open refinement $\mathcal{V} = \{V_\alpha\}_{\alpha \in \Lambda}$ such that each $\emptyset \neq V_\alpha^- \subset U_\alpha$.

   *Sketch of Proof.* It is easily seen that (a) implies (b) since, for disjoint closed subsets $A$ and $B$ of $X$, $\{X - A, X - B\}$ is an open locally finite cover of $X$.

   (b) implies (a). Let $\preceq$ be a well-order for $\Lambda$ and let 0 denote the first element of $\Lambda$ (with respect to $\preceq$). We will use the Transfinite Induction Theorem (see 0.20). Let $L_0 = X - \bigcup\{U_\alpha | \alpha \in \Lambda, \alpha \neq 0\}$. $L_0$ is a closed (maybe $L_0 = \emptyset$) subset of $U_0$. Pick nonempty open set $H_0 \subset X$ such that $L_0 \subset H_0 \subset H_0^- \subset U_0$. Then $\{H_0\} \cup \{U_\alpha | \alpha \in \Lambda, \alpha \neq 0\}$ is an open cover of $X$. Assume that we have defined nonempty open sets $H_\alpha \subset X$, for all $\alpha \prec$ some $\beta$, such that

(i) $\bigcup_{\sigma \preceq \alpha} H_\sigma \cup \bigcup_{\alpha \preceq \sigma} U_\sigma = X,$

(ii) $H_\alpha \subset H_\alpha^- \subset U_\alpha.$

Let $L_\beta = X - (\bigcup_{\sigma \preceq \beta} H_\alpha \cup \bigcup_{\beta \preceq \alpha} U_\alpha)$. $L_\beta$ is a closed subset of $U_\beta$ (why?). Pick a nonempty open set $H_\beta$ such that $L_\beta \subset H_\beta \subset H_\beta^- \subset U_\beta$.

By Transfinite induction (see 0.20), we can then find nonempty open $H_v$, $v \in \Lambda$, which satisfy (i) and (ii) above. It remains to prove that $\{H_v\}_{v \in \Lambda}$ covers $X$. Say $x \in X$ is an element of $U_{\alpha_1}, \ldots, U_{\alpha_n}$, only. (Why can we say this? Have we used local finiteness or something weaker?). Let $\alpha_0 = \max\{\alpha_1, \ldots, \alpha_n\}$. Then $x \in X - \bigcup_{\alpha \prec \alpha_0} U_\alpha$; hence $x \in \bigcup_{\alpha \leq \alpha_0} H_\alpha$ (why?).

12. Let $Y = (E^2, \gamma_s)$ be the Sorgenfrey plane (see ex. 1.7). Show that

    (a) $Y$ is separable (see ex. 1.28),

    (b) $Y$ is not Lindelöf. (Hint: Clearly the set $A = \{(x,y) \in Y | y = -x\}$ is closed in $Y$. Let $\mathcal{U} = \{Y - A\} \cup \{[x, x+1[x[y, y+1[\,|y = -x\}$. $\mathcal{U}$ has no countable subcover!)

    (c) The product of Lindelöf spaces may fail to be Lindelöf. (Hint: Note that $Y = X \times X$, where $X$ is the Sorgenfrey line.)

    (d) $Y$ is not paracompact (see ex. 6). Therefore, the product of paracompact spaces may fail to be paracompact (see ex. 7). As a matter of fact, $Y$ is not normal but this is much harder to see.

13. Let $X$ be Hausdorff. Suppose there exists a countable open cover $\{U_n | n \in \mathbf{N}\}$ of $X$ such that each $U_n^-$ is compact and $U_n^- \subset U_{n+1}$. Show that $X$ is paracompact. (Hint: Use Theorem 8(b).)

14. Let $X$ be a paracompact space and $Y$ a compact Hausdorff space. Show that $X \times Y$ is paracompact. (Hint: The key idea and picture appear in the proof of Theorem 3.8.)

15. Is every locally compact Hausdorff space paracompact (see ex. 8)?

16. A normal space $X$ is called *perfectly normal* if each closed subset $A$ of $X$ is the intersection of a countably many open subsets of $X$ (*i.e.*, $A$ is a $G_\delta$-set); equivalently, each open subset $U$ of $X$ is the union of countably many closed subsets of $X$ (*i.e.*, $U$ is an $F_\sigma$-set). Prove that if $U$ is an open $F_\sigma$-subset of $X$ then there exists a continuous function $f : X \to I$ such that $f^{-1}(0) = X - U$. (Hint: Say $U = \bigcup_{n=1}^\infty A_n$ with each $A_n$ closed. By Urysohn's Lemma, pick continuous functions $f_n : X \to I$ such that $f_n(A_n) = 1$ and $f_n(X - U) = 0$. Finally let $f(x) = \sum_{n=1}^\infty 2^{-n} f_n(x)$, for each $x \in X$. Then

    (i) $0 \leq f(x) \leq 1$, for each $x \in X$,

    (ii) $f$ is continuous,

    (iii) $f^{-1}(0) = X - U$, since $x \in U$ implies $x \in$ some $A_n$ and, hence, $f_n(x) = 1$.

17. Show that the space of ex. 1.27 is not perfectly normal. Indeed, show that the set $\{p\}$ is not a $G_\delta$-set.

18. Show that every metric space $(X, d)$ has a $\sigma$-discrete base (*i.e.*, a base which is a $\sigma$-discrete cover of $X$). (Hint: For each $n$, let $\mathcal{U}_n = \{B(x, \frac{1}{n}) | x \in X\}$ and let $\mathcal{B}_n$ be a $\sigma$-discrete open refinement of $\mathcal{U}_n$; see the proof of Theorem 9. Then $\mathcal{B} = \bigcup_{n=1}^\infty \mathcal{B}_n$ is a $\sigma$-discrete base for $X$.)

19. Show that a *regular space $X$ is metrizable iff $X$ has a $\sigma$-discrete base*.

*Sketch of a Proof.* The *only if* part follows from ex. 18.

For the *if* part there are many steps. Let $\mathcal{B} = \bigcup_n \mathcal{B}_n$ be a $\sigma$-discrete base for $X$.

(1) $X$ is paracompact (hence normal): Every open cover $\mathcal{U}$ of $X$ has a refinement $\mathcal{B}' = \bigcup_n \mathcal{B}'_n$ with $\mathcal{B}' \subset \mathcal{B}$. Now use Theorem 8(b).

(2) $X$ is perfectly normal. Let $U$ open in $X$. For each $n$, let $A_n = \bigcup\{B^- | B \in \mathcal{B}_n$ and $B^- \subset U\}$. Then each $A_n$ is closed, by Lemma 6(b), and $U = \bigcup_{n=1}^\infty A_n$ ($X$ is regular!).

(3) For each $B \in \mathcal{B}_n$ and $n \in \mathbf{N}$, let $f_{Bn} : X \to I$ be a continuous function such that $f_{Bn}^{-1}(0) = X - B$ (see ex. 16). Then, for all $x$, $y \in X$, let
$$\rho_n(x, y) = \sup_{B \in \mathcal{B}_m} |f_B(x) - f_B(y)|$$
and
$$\rho(x, y) = \sum_n 2^{-n} \rho_n(x, y).$$

(Note that, for each $x$, $y \in X$ and $n \in \mathbf{N}$, there exists at most two elements $B_x, B_y \in \mathcal{B}_n$ to which $x$ and $y$ may belong. Then, $\rho_n(x, y) = \sup\{|f_{B_x}(x) - f_{B_x}(y)|, |f_{B_y}(x) - f_{B_x}(y)|\}$.) It is easily checked that $\rho$ is a metric. (Indeed each $\rho_n$ satisfies all properties of a metric, except that $\rho_n(x, y) = 0$ does not necessarily imply $x = y$; furthermore, each $\rho_n \leq 1$.)

(4) Each ball $B(x, \varepsilon)$ is the union of elements of $\mathcal{B}$: Let $y \in B(x, \varepsilon)$. Say $\rho(x, y) = \mu$ and $0 < \delta = \varepsilon - \mu$. Since $\sum_{n=k}^\infty 2^{-n} \rho(w, z) \leq \sum_{n=k}^\infty 2^{-n} = 2^{-n+1}$, for all $w$, $z \in X$, there exists integer $m$ such that $\sum_{n=m}^\infty 2^{-n} \rho_n(w, z) < \frac{\delta}{2}$, for all $w$, $z \in X$. For $j = 1, \ldots, m - 1$, pick elements $B_1, \ldots, B_{m-1}$ of $\mathcal{B}$ ($B_i$ not necessarily in $\mathcal{B}_i$) such that $\rho_j(y, z) < \frac{\delta}{2}$, for all $z \in B_j$, $j = 1, \ldots, m - 1$ (use the continuity of the functions $f_{Bj}$ and the discreteness of each $\mathcal{B}_j$; note that $y$ is in at most one $B^-$ such that $B \in \mathcal{B}_j$).

Then, letting $B = B_1 \cap \cdots \cap B_{m-1}$, we get that, for each $z \in B$, $\rho(x, z) \leq \rho(x, y) + [\sum_{j=1}^{m-1} 2^{-j} \rho_j(x, z) + \sum_{n=m}^\infty 2^{-n} \rho_n(x, z)] < \varepsilon$; that is, $x \in B \subset B(x, \varepsilon)$ with $B \in \mathcal{B}$.

(5) Each $B \in \mathcal{B}$ is a union of $\rho$-balls: Let $B \in \mathcal{B}_k$ and pick $x \in B$. Then, $f_{Bk}(x) = \varepsilon > 0$. Show that $B(x, 2^{-k}\varepsilon) \subset B$ (observe that $\rho_k(x, y) < \varepsilon$ implies that $y \in B$ and $\rho(x, y) < 2^{-k}\varepsilon$ implies that $\rho_k(x, y) < \varepsilon$).

To complete the proof, apply Lemma 1.20.

20. Show that a *regular space* $X$ is *metrizable iff* $X$ *has* a *$\sigma$-locally finite* base. (Hint: Follow the method of proof of ex. 19, exercising a bit more care for the *if* part.)

21. Show that, *for a regular space $X$, the following are equivalent:*
    (a) $X$ *is metrizable.*
    (b) $X$ *has a $\sigma$-discrete base.*
    (c) $X$ *has a $\sigma$-locally finite base.*
    (Hint: See ex. 19 and 20.)

22. A cover $\mathcal{U}$ of a space $X$ is called *closure-preserving* if, for each $\mathcal{V} \subset \mathcal{U}, \bigcup\{V^-|V \in \mathcal{V}\}$. Show that $X$ is paracompact implies that every open cover of $X$ has an open closure-preserving refinement. (The converse is true but the proof is horrendous.)

23. Let $\mathcal{U}$ and $\mathcal{V}$ be covers of a space $X$. $\mathcal{U}$ is a $\Delta$-*refinement* of $\mathcal{V}$ if, for each $x \in X$, $\bigcup\{U \in \mathcal{U}|x \in U\} \subset$ some $V \in \mathcal{V}(\bigcup\{U \in \mathcal{U}|x \in U\}$ is generally denoted by $st(x,\mathcal{U})$ and called the *star of $x$* with respect to $\mathcal{U}$). Show that $X$ is paracompact implies that every open cover of $X$ has an open $\Delta$-refinement. (The converse is true but the proof is quite difficult.) (*Sketch of a proof.* Let $\mathcal{U}$ be an open cover of $X$, $\mathcal{V}$ an open locally finite refinement of $\mathcal{U}$, $\mathcal{V}'$ a closed cover of $X$ such that $\bar{V}' \subset V$, for each $V \in \mathcal{V}$ (see Lemma 11). For each finite subcollection $\mathcal{F} = \{V_1, \ldots, V_n\}$ of $\mathcal{V}$, let

$$w(\mathcal{F}) = (V_1 \cap \cdots \cap V_n) - \bigcup\{\bar{V}'|V \notin \mathcal{F}\}.$$

Let $\mathcal{W} = \{w(\mathcal{F})|\mathcal{F} \text{ is a finite subcollection of } \mathcal{V}\}$. Then
    (i) $\mathcal{W}$ is an open cover of $X$.
    (ii) $x \in V_0' \subset \bar{V}_0' \subset V_0$ implies $st(x,\mathcal{W}) \subset V_0$,
    (iii) $\mathcal{W}$ is locally finite.)

24. (Tietze's Extension Theorem.) A space $X$ is normal iff, whenever $A$ is a closed subset of $X$ and $f : A \to I$ is continuous, there exists a continuous $\bar{f} : X \to I$ such that $\bar{f}|A = f$ (*i.e.*, $\bar{f}$ is a *continuous extension* of $f$ to all of $X$). (Hint: The *if* part is obvious since, given disjoint closed subsets $C$ and $B$ of $X$, the function $f : C \cup B \to I$ such that $f(C) = 0$ and $f(B) = 1$ is continuous. Therefore, .... The proof of the *only if* part is hard: Let $A_1 = \{x \in A|f(x) \le \frac{1}{3}\}$ and $B_1 = \{x \in A|f(x) \le \frac{2}{3}\}$. Applying Urysohn's Lemma (see ex. 4.9), there exists a continuous $f_1 : X \to [\frac{1}{3}, \frac{2}{3}]$ such that $f_1(A_1) = \frac{1}{3}$ and $f_1(B_1) = \frac{3}{2}$. Clearly, for each $a \in A$, $|f(a) - f_1(a)| \le \frac{1}{3}$ and, hence, $g_1 = f - f_1$ maps $A$ to $[0, \frac{1}{3}]$. Repeating the process of removing middle thirds, with $g_1$ instead of $f$ and $[0, \frac{1}{3}]$ instead of $[0,1]$, let $A_2 = \{x \in A|g_1(x) \le \frac{1}{9}\}$, $B_2 = \{x \in A|g_1(x) \le \frac{2}{9}\}$ and find $f_2 : X \to [\frac{1}{9}, \frac{2}{9}]$ such that $f_2(A_2) = \frac{1}{9}$, $f_2(B_2) = \frac{2}{9}$. Clearly $|(f - f_1) - f_2| = |f - (f_1 + f_2)| < (\frac{1}{3})^2$ on $A$.

Inductively, one then obtains continuous functions $f_i : X \to [0, \frac{1}{3^n}] \subset [0,1]$ such that $|f(a) - \sum_{k=1}^{n} f_k(a)| \le (\frac{1}{3})^n$, for all $a \in A$ (*i.e.*, $\{\sum_{k=1}^{n} f_k\}_n$ converges uniformly to $f$ on $A$).

Define $\bar{f} : X \to E^1$ by $\bar{f}(x) = \sum_{t=1}^{\infty} f_i(x)$, and show that

(a) $0 \le \bar{f}(x) \le 1$, for each $x \in X$ (geometric series!),

(b) $\bar{f}(a) = f(a)$, for each $a \in A$,

(c) $\bar{f}$ is continuous. (For $x \in X$ and $\varepsilon > 0$, pick integer $m$ such that $\sum_{n=m+1}^{\infty} (\frac{1}{3})^n < \frac{\varepsilon}{2}$. Pick neighborhoods $U_i$ of $x$, for $i = 1, \ldots, m$, such that $y \in U$, implies $|f_i(x) - f_i(y)| < \frac{\varepsilon}{2m}$. Let $U = U_1 \cap \ldots \cap U_m$ and show that $y \in U \Rightarrow |\bar{f}(x) - \bar{f}(y)| < \varepsilon$.)

25. (Variations on Tietze's Extension Theorem.) Show that Tietze's Extension Theorem remains valid if

(a) $I$ is replaced by any closed interval $[a, b]$. (Hint: Use the homeomorphism $h : I \to [a, b]$, defined by $h(t) = ta + (1 - t)b$.)

(b) $I$ is replaced by any open interval $]a, b[$. (Hint: Suffices to consider $f : A \to ]-1, 1[ \subset [-1, 1]$ (why?). From (a) extend $f$ to $\bar{f} : X \to [-1, 1]$. Let $A_0 = \{x \in X | \bar{f}(x) = -1 \text{ or } 1\}$. Clearly $A$ and $A_0$ are disjoint closed subsets of $X$. Pick continuous $g : X \to I$ such that $g(A_0) = \frac{1}{3}$ and $g(A) = 1$ (why?). Define $\tilde{f} : X \to I$ by $\tilde{f}(x) = g(x)\bar{f}(x)$. Show that $\tilde{f} : X \to ]-1, 1[$, $\tilde{f}$ is continuous and $\tilde{f}|A = f$.)

# Chapter 8

# The Fundamental Group

It is intuitively obvious that no amount of stretching, shrinking and deforming, *without tearing or gluing*, will transform a closed disc into an annulus; that is, the annulus and the closed disc are not homeomorphic. The same comments apply to the 2-sphere and the torus. Yet, these highly intuitive facts are equally difficult to prove; both pairs of spaces are compact metrizable, connected, locally connected and arcwise connected; both pairs are even locally homeomorphic.

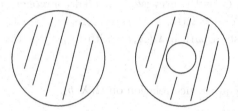

In the search for a proof that these pairs of spaces, and many other pairs, are not homeomorphic, it is interesting to observe that any two *rubber bands* laid out on the surface of the closed disc or the sphere can be *continuously deformed* into each other by stretching, shrinking, bending and *gluing*, without tearing. The same cannot be said about the annulus or the torus: Consider a rubber band laid on the inner rim of the annulus and one that does not surround this rim; also consider a rubber band around the *outer hole* of the torus and one around the *inner hole* of the torus.

Let us give this simple observation precise mathematical form and reap some of its many powerful benefits.

## 8.1 Description of $\prod_1(X, b)$

**1. Definition.** Let $X$ be a topological space and let $b$, $c$, $d \in X$.

(a) A *loop based at b* is a path $\alpha : I \to X$ such that $\alpha(0) = b = \alpha(1)$. (Here is the rubber band!)

(b) Two paths $\alpha$ and $\beta$ from $c$ to $d$ are *homotopic* (i.e., $\alpha \underset{cd}{\sim} \beta$) provided that there exists a continuous map $h : I \times I \to X$ such that $h(t,0) = \alpha(t)$; $h(t,1) = \beta(t)$, $h(0,t) = c$, $h(1,t) = d$, for $t \in I$. (Here is the *continuous deformation!*). The map $h$ is called a *homotopy from $\alpha$ to $\beta$*. If $c = d$, $\alpha \underset{cd}{\sim} \beta$ becomes $\alpha \underset{cd}{\sim} \beta$.

(c) Let $\alpha$ be a path from $x_0$ to $x_1$ and $\beta$ be a path from $x_1$ to $x_2$. Then the product of $\alpha$ and $\beta$ is the path $\alpha * \beta$ from $x_0$ to $x_2$, defined by

$$\alpha * \beta(t) \begin{cases} \alpha(2r), & 0 \le t \le 1/2, \\ \beta(2t-1), & 1/2 \le t \le 1. \end{cases}$$

(Clearly, $\alpha * \beta$ is well-defined and continuous!)

(d) The *reverse* of the path $\alpha$ is the path $\alpha^r$, defined by $\alpha^r(t) = \alpha(1-t)$.

(e) $\Omega(X,b)$ denotes the set of all loops on $X$ based at $b$.

Note that the product of two paths amounts to no more than *travelling through both with twice the original velocity*, while the reverse of a path is no more than *travelling on the same path in the opposite direction*. Definitely, the reverse of a path is not related to the concept of the inverse function.

Also, note that if $C_b$ is the arcwise connected component of $X$ which contains the point $b$, then $\Omega(X,b) = \Omega(C_b,b)$; furthermore, $\alpha$, $\beta \in \Omega(X,b)$ are homotopic *if and only if* there exists a path in $\Omega(X,b)$, with respect to the co topology, from $\alpha$ to $\beta$ (see ex. 1).

**2. Lemma.** $\underset{b}{\sim}$ is an equivalence relation on $\Omega(X,b)$.

**Proof.** Clearly, $\alpha \underset{b}{\sim} \alpha$, for each $\alpha \in \Omega(X,b)$. Say $\alpha \underset{b}{\sim} \beta$. Let $h$ be a homotopy from $\alpha$ to $\beta$. Letting $h'(s,t) = h(s,1-t)$, for each $(s,t) \in I \times I$, we immediately get that $h'$ is a homotopy from $\beta$ to $\alpha$. Therefore, $\alpha \underset{b}{\sim} \beta$ iff $\beta \underset{b}{\sim} \alpha$. Finally, suppose $\alpha \underset{b}{\sim} \beta$ and $\beta \underset{b}{\sim} \gamma$. Say $h_1$ is a homotopy from $\alpha$ to $\beta$ and $h_2$ is a homotopy from $\beta$ to $\gamma$. Define $h : I \times I \to X$ by

$$h(s,t) = \begin{cases} h_1(s,2t), & 0 \le t \le 1/2, \\ h_2(s,2t-1), & 1/2 \le t \le 1. \end{cases}$$

It is clear that $h$ is a homotopy from $\alpha$ to $\gamma$; therefore $\alpha \underset{b}{\sim} \beta$, $\beta \underset{b}{\sim} \gamma$ implies $\alpha \underset{b}{\sim} \gamma$.

Let $\prod_1(Y,b)$ denote the set of $\underset{b}{\sim}$-equivalence classes of $\Omega(Y,b)$ and define an operation $\bullet$ on this set by

$$[\alpha] \bullet [\beta] = [\alpha * \beta]$$

($[\gamma]$ denotes the equivalence class of the loop $\gamma$).

We will now show that the operation $\bullet$ on $\prod_1(Y, b)$ is well-defined and makes $\prod_1(Y, b)$ into an algebraic group. Hence forth, let $[\alpha] \bullet [\beta] = [\alpha][\beta]$.

**3. Lemma.** The operation $\bullet$ on $\prod_1(Y, b)$ is well-defined (*i.e.*, if $[F] = [f]$ and $[G] = [g]$ then $[f * g] = [F * G]$).

**Proof.** Suffices to show that if $F \underset{b}{\simeq} f$ and $G \underset{b}{\simeq} g$ then $F * G \underset{b}{\simeq} f * g$. So, let $h_1$, $h_2$ be homotopies such that $h_1(x, 0) = F(x)$, $h_1(x, 1) = f(x) h_1(0, t) = b = h_1(1, t) h_2(x, 0) = G(x)$, $h_2(x, 1) = g(x) h_2(0, t) = b = h_2(1, t)$, and define $h : I \times I = Y$ by

$$h(x, t) = \begin{cases} h_1(2x, t), & 0 \le x \le 1/2 \\ h_2(2x - 1, t), & 1/2 \le x \le 1. \end{cases}$$

It is easily seen that $h$ is a homotopy between $F * G$ and $f * g$. Hence "$\bullet$" is well-defined and single-valued.

**4. Lemma.** The operation "$\bullet$" on $\prod_1(Y, b)$ is associative.

**Proof.** Clearly, it suffices to prove that $(f * g) * h \underset{b}{\simeq} f * (g * h)$ for any $f, g, h \in \Omega(Y, b)$. Note that

$$(f * g) * h(x) = \begin{cases} f(4x), & 0 \le x \le 1/4, \\ g(4x - 1), & 1/4 \le x \le 1/2, \\ h(2x - 1), & 1/2 \ge x \le 1, \end{cases}$$

$$f * (g * h)(x) = \begin{cases} f(2x), & 0 \le x \le 1/2, \\ g(4x - 2), & 1/2 \le x \le 3/4, \\ h(4x - 3), & 3/4 \ge x \le 1. \end{cases}$$

Now, we define $H : I \times I \to Y$ by

$$H(x, t) = \begin{cases} f\left(\frac{4x}{t+1}\right), & t \le 4x - 1, \\ g(4x - t - 1), & 4x - l \ge t \ge 4x - 2, \\ h\left(\frac{4x - t - 2}{2 - t}\right), & 4x - 2 \ge t. \end{cases}$$

Note that $H$ is constant along any segment of line in the middle strip which is parallel to the line $t = 4x - 1$.

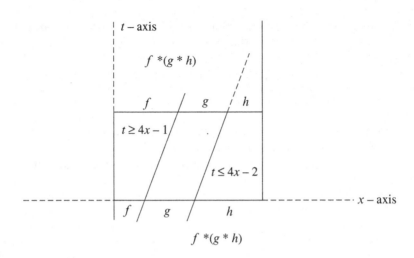

$$f * (g * h)$$

It is then easily (but tediously) seen that $H$ is a homotopy between $(f * g) * h$ and $f * (g * h)$.

**5. Lemma.** If $c_b$ is the constant map from $I$ to $b$ then $[c_b]$ is the identity element of $\prod_1(Y, b)$.

**Proof.** It suffices to prove that, for each $f \in \Omega(Y, b)$, $f * c_b \underset{\tilde{b}}{\sim} f$. So, define $H : I \times I \to Y$ by

$$H(x, t) \begin{cases} f\left(\frac{2x}{1+t}\right), & t \geq 2x - l, \\ b, & t \leq 2x - 1. \end{cases}$$

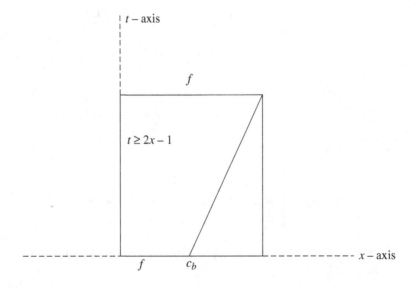

**6. Lemma.** If $[f] \in \prod_1(Y, b)$ then $[f] \bullet [f^r] = [c_b]$ (*i.e.*, each element of $\prod_1(Y, b)$ has an inverse in $\prod_1(Y, b)$ with respect to the operation "$\bullet$").

**Proof.** It suffices to show that $f * f^r \underset{b}{\simeq} c_b$, where

$$f * f^r(x) = \begin{cases} f(2x), & 0 \le x \le 1/2, \\ f^r(2x - 1) = f(2 - 2x), & 1/2 \le x \le 1. \end{cases}$$

Simply define $H : I \times I \to Y$ by

$$H(x, t) = \begin{cases} f(2x), & t \le 1 - 2x \text{ and } 0 \le x \le 1/2, \\ f(1 - t) = g(t), & t \ge 1 - 2x \text{ and } 0 \le x \le 1/2, \\ f(2 - 2x), & t \le 2x - 1 \text{ and } 1/2 \le x \le 1. \end{cases}$$

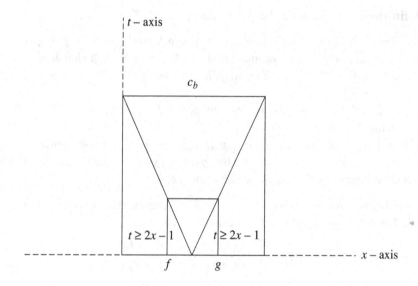

**7. Theorem.** For any space $Y$ and $b \in Y$, $(\prod_1(Y, b), \bullet)$ is a group.

**Proof.** Immediate from Lemmas 3 through 6.

The group $\prod_1(Y, b)$ is called the *first homotopy group* (or the *fundamental group*) of $Y$ with base point $b$. Fortunately, for a very large class of spaces, different base points yield isomorphic fundamental groups. Later on (see ex. 2) we will see that the next result is best possible.

**8. Lemma.** Let $b, c \in X$ and $\gamma$ be an arc from $b$ to $c$. Then $\prod_1(X, b) \approx \prod_1(X, c)$.

**Proof.** Define $\gamma_\# : \prod_1(X, b) \to \prod_1(X, c)$ by $\gamma_\#([\alpha]) = [\gamma^r * \alpha * \gamma]$. Note that

$$\gamma_\#([\alpha * \beta]) = [\gamma^r * \alpha * \beta * \gamma] = [\gamma^r * \alpha * \gamma * \gamma^r * \beta * \gamma]$$
$$= [\gamma^r * \alpha * \gamma] \bullet [\gamma^r * \beta * \gamma] = \gamma_\#([\alpha]) \bullet \gamma_\#([\beta]),$$

which shows that $\gamma_\#$ is a homomorphism. Clearly $\gamma_\#$ is one-to-one and onto (indeed, $(\gamma_\#)^{-1} = (\gamma')_\#$); hence $\gamma_\#$ is an isomorphism.

Lemma 8 tells us that each arcwise connected space has one and only one (up to isomorphism) fundamental group and that to study the fundamental groups of any space it suffices to study the fundamental groups of each arcwise connected component of that space. For arcwise connected spaces $X$, it is customary to let $\prod_1(X, b) = \prod_1(X)$, since the fundamental group does not depend on the base point.

## 8.2   Elementary Facts about $\prod_1(X, b)$

The justification for the following definition will soon become apparent.

**9. Definition.** Let $X$ and $Y$ be spaces and pick $b \in X$.

(a) Two continuous maps $f$, $g : X \to Y$ are *homotopic* (i.e., $f \sim g$) provided that there exists a continuous map $h : X \times I \to Y$ such that $h|X \times \{0\} = f$ and $h|X \times \{1\} = g$. The map $h$ is called a *homotopy from $f$ to $g$* (i.e., $h : f \sim g$).

(b) If $f : X \to Y$ is a continuous map, let $f_* : \prod_1(X, b) \to \prod_1(Y, f(b))$ be defined by $f_*([\alpha]) = [f \circ \alpha]$.

(c) $X$ and $Y$ are of the *same homotopy type* if there exist continuous maps $f : X \to Y$, $g : Y \to X$ such that $g \circ f \sim i_X$ and $f \circ g \sim i_Y$. (See 0.10)

(d) $X$ is *contractible* if $i_X \sim c_b$, for some $b \in X$.

It is obvious that $(i_X)_* = i_{\prod_1(X,b)}$. Also, a contractible space is arcwise connected: Let $h : i_X \sim c_b$, and $c$, $d \in X$; define $\alpha : I \to X$ by

$$\alpha(t) = \begin{cases} h(c, 2t), & t \leq 1/2, \\ h(d, 2(1 - t)), & 1 \geq 1/2. \end{cases}$$

**10.   Lemma.** The function $f_* : \prod_1(X, b) \to \prod_1(Y, f(b))$ of Definition 9(b) is a homomorphism.

**Proof.** Note that, for $[\alpha]$, $[\beta] \in \prod_1(X, b)$,

$$f_*([\alpha][\beta]) = f_*([\alpha * \beta]) = [f \circ (\alpha * \beta)] = [(f \circ \alpha) * (f \circ \beta)]$$
$$= [f \circ \alpha][f \circ \beta] = f_*([\alpha])f_*([\beta]).$$

**11. Theorem.** Let $X$, $Y$, $Z$ be arcwise connected spaces and pick $b \in X$; let $f$, $g : X \to Y$ and $k : Y \to Z$ be continuous maps. Then

(a) $(k \circ f)_* = k_* \circ f_*$.

(b) If $h : f \sim g$ then $g_* = \sigma_\# \circ f_*$, where $\sigma = h|\{b\} \times I$ is the path from $f(b)$ to $g(b)$. (See proof of Lemma 8.)

(c) If $X$ and $Y$ are of the same homotopy type then $\prod_1(X) \approx \prod_1(Y)$.

**Proof.** Part (a). Let $[\alpha] \in \prod_1(X, b)$. Then

$$(k \circ f)_*([\alpha]) = [k \circ f \circ \alpha] = k_*([f \circ \alpha]) = k_* \circ f_*([\alpha]).$$

Part (b). Let $[\alpha] \in \prod_1(X, b)$. Note that $g_*([\alpha]) = [g \circ \alpha]$ and $\sigma_\# \circ f_*([\alpha]) = [\sigma' * f \circ \alpha * \sigma]$; therefore, we must show that $\sigma' * f \circ \alpha * \sigma \underset{g(b)}{\sim} g \circ \alpha$: First, define $h_1 : I \times I \to Y$ by $h_1(s,t) = h(\alpha(s), t)$, and note that $h_1 : f \circ \alpha \sim g \circ \alpha$ such that $h_1|\{0\} \times I = \sigma = h_1|\{1\} \times I$. From the diagram

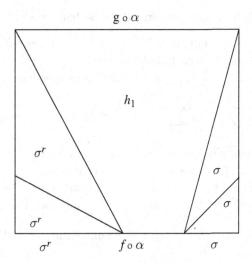

it is easy to obtain an analytic expression for a homotopy $h_2 : \sigma^r * f \circ \alpha * \sigma \underset{g(b)}{\sim} g \circ \alpha$ (see ex. 3).

Part (c). Let $f : X \to Y$ and $g : Y \to X$ be continuous maps such that $i_X \sim g \circ f$ and also $i_Y \sim f \circ g$. It follows that $f_* : \prod_1(X, b) \to \prod_1(Y, f(b))$ and $g_* : \prod_1(Y, f(b)) \to \prod_1(X, gf(b))$; from (a) and proof of Lemma 8, $g_* \circ f_* = (g \circ f)_* = \sigma_\# \circ (i_X)_* = \sigma_\#$, $f_* \circ g_* = (f \circ g)_* = \gamma_\# \circ (i_Y)_* = \gamma_\#$, where $\sigma_\#$ and $\gamma_\#$ are isomorphisms; therefore, $g_* \circ f_* = \sigma_\#$ implies that $g_*$ is onto and $f_*$ is $1-1$, while $f_* \circ g_* = \gamma_\#$ implies that $g_*$ is $1-1$ and $f_*$ is onto, which shows that $f_*$ and $g_*$ are isomorphisms. From Lemma 8 we get that $\prod_1(X) \approx \prod_1(Y)$.

**12. Corollary.** The following is true.

(a) If $X$ is contractible then $\prod_1(X) = 1$ (*i.e.*, the trivial group consisting of the unit element only, the *double* meaning of 1 will be clear from the context).

(b) $\prod_1(E^n, 0) = 1$.

**Proof.** Part (a) follows immediately from Theorem 11(c), since $\prod_1(\{b\}, b) = 1$, for any singleton $\{b\}$. Part (b) follows immediately from (a), since $E^n$ is contractible (note that $h : E^n \times I \to E^n$, defined by $h(x,t) = tx$, is a homotopy from the identity map $i_X$ to the constant map $c_0$).

## 8.3 Simplicial Complexes

While it is easy to define the fundamental group of a space it is extremely difficult to determine the structure of that group, unless that space is a *nice* union of *nice* subspaces. Fortunately, most common spaces are in this category which we are about to study.

**13. Definition.** To minimize clutter, we use no bars on vertices.

(a) A set $\{v_0, v_1, \ldots, v_k\} \subset E^n$ is said to be *convex-independent* if $\{v_1 - v_0, \ldots, v_k - v_0\}$ is linearly independent.

(b) Suppose $\{v_0, v_1, \ldots, v_k\} \subset E^n$ is convex-independent. Then the set

$$\left\{ \sum_{i=0}^{k} \lambda_1 v_i \,\middle|\, \sum_{i=0}^{k} \lambda_i = 1, \ 0 < \lambda_i < 1, \ \text{for } i = 0, \ldots, k \right\}$$

is called the *open $k$-simplex*, with vertices $v_0, \ldots, v_k$, and is denoted by $\langle v_0, \ldots, v_k \rangle$. The *closed $k$-simplex*, with vertices $v_0, \ldots, v_k$, consists of $\langle v_0, \ldots, v_k \rangle$ with its boundary in $E^n$ and is denoted by $\langle v_0, \ldots, v_k \rangle^-$. *Unless otherwise stated*, we will let $s^n, \ldots$ denote a closed $n$-simplex. Let $\langle v \rangle$ denote a 0-simplex with vertex $v$.

**Remark.** It is obvious that $k \leq n + 1$ and easily seen that $\langle v_0, \ldots, v_k \rangle^- = \{\sum_{i=0}^{k} \lambda_i v_i | \sum_{i=0}^{k} \lambda_i = 1, \ 0 \leq \lambda_i \leq 1, \ \text{for } i = 0, \ldots, k\}$; furthermore, for each $\sum_{i=0}^{k} \lambda_i v_i$, $\sum_{i=0}^{k} \beta_i v_i \in \langle v_0, \ldots, v_k \rangle$ and $t \in I$, $t \sum_{i=0}^{k} \lambda_i v_i + (1-t) \sum_{i=0}^{k} \beta_i v_i = \sum_{i=0}^{k} [t\lambda_i + (1-t)\beta_i] v_i \in \langle v_0, \ldots, v_k \rangle$; that is, $\langle v_0, \ldots, v_k \rangle$ is a convex subset of $E^n$. Similarly, $\langle v_0, \ldots, v_k \rangle^-$ is a convex subset of $E^n$. Also $\sum_{i=0}^{k} \lambda_i v_i = \sum_{i=0}^{k} \beta_i v_i$ iff $\lambda_0 = \beta_0, \ldots, \lambda_k = \beta_k$. (The *if* part is obvious and the *only if* part goes as follows: Note that

$$0 = \sum_{i=0}^{k} (\lambda_i - \beta_i) v_i = \sum_{i=0}^{k} (\lambda_i - \beta_i) v_i - \left( \sum_{i=0}^{k} \lambda_i - \sum_{i=0}^{k} \beta_i \right) v_0$$

$$= \sum_{i=0}^{k} (\lambda_i - \beta_i)(v_i - v_0).$$

Since $\{v_1 - v_0, \ldots, v_k - v_0\}$ is linearly independent, we then get that $\lambda_1 - \beta_1 = 0, \ldots, \lambda_k - \beta_k = 0$, which does the trick.)

**14. Definition.** A *simplicial complex* (or *polytope*) $K$ is a space which satisfies the following:

(i) $K = \bigcup_{\alpha \in \Lambda} s_\alpha$ such that each $s_\alpha$ is a closed simplex.

(ii) For every $\alpha, \beta \in \Lambda$, $s_\alpha \cap s_\beta$ is a closed simplex.

**Remark.** Let $\{v_\mu\}_{\mu\in\Gamma}$ be the vertices of the polytope $K$. Then, from the preceding remark; $x \in K$ implies $x = \sum_\mu x_\mu v_\mu$, uniquely, with only finitely many $x_\mu \neq 0$. $(x_\mu)_{\mu\in\Gamma}$ are called the *barycentric coordinates* of $x$ in $K$.

**Remark.** Condition (i) of the preceding definition imposes severe restrictions on the topology that $K$ can have. One possible and extremely useful topology is the *weak topology* which is defined as follows: A set $U \subset K$ is open provided that $U \cap s_\alpha$ is open in $s_\alpha$, for each $\alpha \in \Lambda$. (It is easy to check that this, indeed, defines a topology on $K$ such that each $s_\alpha$ is a subspace of $K$, because of (ii) of Definition 14; without it we would have nonsense. See ex. 4.)

**15. Definition.** A simplicial complex $K$ with the weak topology is called a *CW*-polytope. Henceforth, *polytopes will be assumed to have the CW-topology whenever a topology is required.*

**16. Definition.**

(a) A polytope $K$ with only finitely many simplexes will be called a finite polytope.

(b) $K'$ is called a subpolytope of $K$ if $K'$ is a polytope and $K'$ is a subspace of $K$.

**17. Definition.**

(a) Let $K$ be a polytope. For each $n \in \mathbf{N}$, let $K(n)$ be the subpolytope of $K$ which consists of all $j$-simplexes of $K$ for $j = 0, \ldots, n$. $K(n)$ is called the $n$-skeleton of $K$.

(b) A polytope $K$ is said to be $n$-dimensional if $K(n) \neq \emptyset$ and $K(n+i) = K(n)$, for $i \in \mathbf{N}$.

**18. Definition.** A space $X$ is said to be *triangulated* if $X$ is homeomorphic to a polytope.

Clearly, all $n$-spheres and all $n$-balls are triangulated, since (the boundary of) an $n$-ball is homeomorphic to (the boundary polytope of) an $n$-simplex (see ex. 5).

**19. Definition.** Let $K$ be a polytope and $v$ a vertex of $K$. The *star of $v$* is the union of all open simplexes of $K$ having $v$ as a vertex, and it is denoted by $St\, v$. Recall that $\overline{St\, v}$ denotes the closure of $St\, v$ in $K$.

**20. Lemma.** Let $v$ be a vertex of a polytope $K$. Then

(a) $St\, v$ is an open subset of $K$,

(b) $\overline{St\ v}$ is a subpolytope of $K$,

(c) $St\ v = \{x \in K|$ the barycentric $v$-coordinate of $x$ is not zero$\}$,

(d) $\langle v_0, \ldots, v_n \rangle \subset K \cap (\bigcap_{i=0}^n St\ v_i)$ if and only if $\bigcap_{i=0}^n St\ v_i \neq \emptyset$.

**Proof.**

(a) For any closed simplex $g \subset K$, $(K - St\ v) \cap g$ either equals $g$ or a closed face of $g$ or the empty set. Therefore $K - St\ v$ is a closed subset of $K$. (Indeed we have even *shown that $K - St\ v$ is a subpolytope of $K$.)

(b) Obvious. Indeed $\overline{St\ v}$ is the union of all closed simplexes of $K$ having $v$ as a vertex.

(c) Straightforward, by the definitions of $St\ v$ and of *open simplex.*

(d) The *only if* part is obvious. Let us therefore prove the *if* part: Clearly, $\langle V_0, \ldots, v_n \rangle \subset K$, because $\bigcap_{i=0}^n St\ v_i \neq \emptyset$ and $\langle v_0, \ldots, v_n \rangle \subset \bigcap_{i=0}^n St\ v_i$, because of part (c).

**Observation.** The union of any collection of closed simplexes contained in a polytope $K$ is a polytope.

**21. Theorem.** Let $K$ be a polytope. A subset $C$ of $K$ is compact iff $C$ is a closed subset of a finite subpolytope of $K$.

**Proof.** The *if* part is obvious. Let us therefore prove the *only if* part.

Let $C$ be a compact subset of $K$ and suppose $C$ is not contained in a finite subpolytope of $K$. Then there exist finite subpolytopes $K_1 \subset K_2 \subset \ldots$ of $K$ with some $x_n \in (K_n \cap C) - K_{n-1}$, for $n = 2, 3, \ldots$. Then (every subset of) $B = \{x_i | i = 1, 2, \ldots\}$ is a closed subset of $K$, because $K$ has the weak topology over its finite subpolytopes(!). Therefore, the sequence $\{x_i\}_{i=1}^\infty$ has no cluster point, contradicting the fact that $C$ is compact. This completes the proof.

**22. Corollary.** A polytope is compact if and only if it is a finite polytope.

## 8.4   Barycentric Subdivision

**23. Definition.** The point of an $n$-simplex $g^n$ all of whose barycentric coordinates equal $\frac{1}{n+1}$ is called the barycenter of $g^n$ (the barycenter of a 0-simplex $\langle q \rangle$ is $q$).

We will write $g^i < g^n$ if and only if $g^i$ is a proper face of $g^n$ (*i.e.*, $g^i \subset g^n$ and $g^i \neq g^n$).

**24. Lemma.** Let $s_1 < \cdots < s_j$ be proper faces of a simplex $g^n$. Pick $x_1 \in s_1$, and $x_i \in s_i - s_{i-1}$ for $i = 2, \ldots, j$. Then $\{x_1, \ldots, x_j\}$ is convex-independent.

**Proof.** Straightforward.

**25. Definition.** Let $x_1, \ldots, x_k$ be the barycenters of all the faces $s_1, \ldots, s_k$ of a given closed simplex $g^m$. The union of all simplexes $\langle x_{i_0}, \ldots, x_{i_n} \rangle$ such that $s_{i_0} < \cdots < s_{i_n}$ (see Lemma 24) forms a finite polytope which is called the *first barycenteric subdivision* of $g^m$. Inductively, the $n$-fold barycentric subdivision of $g^m$ is the first barycentric subdivision of the $(n-1)$-fold barycentric subdivision of $g^m$. The $n$-fold *barycentric subdivision of any polytope* $K$ is the polytope $K^{(n)}$ which is obtained from $K$ by replacing each closed simplex of $K$ with its $n$-fold barycentric subdivision.

**26. Lemma.** Let $K$ be any polytope. Then $K$ is homeomorphic to $K^{(n)}$, for each $n$.

**Proof.** Clearly, it suffices to prove that $K$ is homeomorphic to $K^{(1)}$. Clearly, each closed simplex $s \subset K$ is homeomorphic to $s^{(1)}$. Therefore, since $K$ and $K^{(l)}$ have the weak topology, one easily sees that $K$ is homeomorphic to $K^{(1)}$ (see ex. 6).

**27. Definition.** The *mesh* of a polytope $K$ is the supremum of the diameters of all simplexes of $K$. (The mesh may be infinite!)

**28. Proposition.** Let $A$ be a subset of $E^n$. Then the diameter of $A$ equals the diameter of the convex hull of $A$ (*i.e.*, $\operatorname{diam} A = \operatorname{diam} \operatorname{conv} A$).

**Proof.** Clearly, $\operatorname{diam} A \leq \operatorname{diam} \operatorname{conv} A$, because $A \subset \operatorname{conv} A$. Let $\operatorname{diam} A = \delta$. Then for any $a \in A$ and $\varepsilon > 0$, $\operatorname{conv} A \subset B(a, \delta + \varepsilon)$, because $A \subset B(a, \delta + \varepsilon)$ and $B(a, \delta + \varepsilon)$ is convex. Therefore, $\operatorname{conv} A \subset \bigcap\{B(a, \delta + \varepsilon) | a \in A, \varepsilon > 0\}$. This easily implies that $\operatorname{diam} \operatorname{conv} A \leq \delta$. Therefore, $\operatorname{diam} \operatorname{conv} A = \delta$.

**29. Corollary.** The diameter of a geometric simplex is the length of its longest 1-face.

**Proof.** Immediate from Proposition 28, since a geometric simplex is the convex hull of the set of its vertices (by its very definition).

**30. Lemma.** Let $g$ be a closed $p$-simplex with diameter $d$. Then mesh $g^{(1)} \leq \frac{pd}{p+1}$.

**Proof.** (By induction). For $p = 0$ we have a valid result. Assume the result is valid for $(n-1)$-simplexes, with $n \geq 1$, and let us show that it is valid for any $n$-simplex $\sigma = \langle x_0, \ldots, x_n \rangle$. By Corollary 29, the induction hypothesis and the fact that $\frac{k}{k+1} \leq \frac{n}{n+1}$, for $k \leq n$, we need only show that the length of any 1-face of $\sigma^{(1)}$ which starts at the barycenter $\sum_{i=0}^{n} \frac{1}{n+1} x_i$ of $\sigma$ and ends at the barycenter

$\sum_{i=0}^{k} \frac{1}{k+1} x_i$ of a $k$-face of $\sigma$ (this may involve a renumbering of the vertices of $\sigma$!) is no larger than $\frac{nd}{n+1}$: Note that

$$\left| \sum_{i=0}^{n} \frac{1}{n+1} x_i - \sum_{i=0}^{k} \frac{1}{k+1} x_i \right| = \left| \sum_{i=0}^{k} \left( \frac{1}{n+1} - \frac{1}{k+1} \right) x_i + \sum_{i=k+1}^{n} \frac{1}{n+1} x_i \right|$$

$$= \frac{1}{(n+1)(k+1)} \left| \sum_{i=0}^{k} (k-n) x_i + \sum_{i=k+1}^{n} (k+1) x_i \right|$$

$$= \frac{1}{(n+1)(k+1)} \left| \sum_{i=k+1}^{n} (k+1) x_i - \sum_{j=0}^{k} (n-k) x_j \right|$$

$$= \frac{1}{(n+1)(k+1)} \left| \text{sum of } (n-k)(k+1) \text{ summands} \right.$$
$$\left. x_i - x_j, n \geq i > k \geq j \geq 0 \right|$$

$$\leq \frac{(n-k)(k+1)d}{(n+1)(k+1)} = \frac{(n-k)d}{n+1} \leq \frac{nd}{n+1}.$$

(Note that we proved more than we claimed—namely, a relationship between the lengths of the 1-faces of $\sigma$ and the 1-faces of $\sigma^{(1)}$.)

**31. Theorem.** Let $K$ be an $n$-dimensional polytope with mesh $K = \lambda < \infty$. Then mesh $K^{(1)} \leq \frac{n\lambda}{n+1}$.

**Proof.** Immediate from Lemma 30.

**32. Theorem.** Let $K$ be an $n$-dimensional polytope. For any $\delta > 0$, there exists $m$ such that mesh $K^{(m)} < \delta$.

**Proof.** Let mesh $K = \lambda$. By Theorem 31 and induction, we get that the mesh of $K^{(m)} \leq \left( \frac{n}{n+1} \right)^m \lambda$ and $\lim_{m} \left( \frac{n}{n+1} \right)^m = 0$. This completes the proof.

## 8.5   Simplicial Approximation

**33. Definition.** Let $K$ and $L$ be polytopes. A map $\psi : K \to L$ is a *simplicial map* provided that

  (1)  $\psi(K(0)) \subset L(0)$.
  (2)  For each $p \in K$, if $p = \sum_i \lambda_i v_i$ then $\psi(p) = \sum_i \lambda_i \psi(v_i)$

(*i.e.*, $\psi$ is linear on each simplex of $K$ and $\psi(K)$ is a subcomplex of $L$).
    Clearly, each simplicial map is continuous.

**34. Definition.** Let $K$ and $L$ be polytopes, and $f : K \to L$ a continuous function.

A simplicial map $\psi : K \to L$ is a *simplicial approximation* to $f$ if $f(St\ v) \subset St\ \psi(v)$, for each $v \in K(0)$.

**35. Lemma.** If $\psi : K \to L$ is a simplicial approximation to $f : K \to L$ then, for each $p \in K$, $f(p)$ and $\psi(p)$ lie in a common closed simplex of $L$.

**Proof.** Pick $v \in K(0)$ such that the barycentric $v$-coordinate of $p$ is not zero. Then, $p \in St\ v$, which implies that $f(p) \in f(St\ v) \subset St\ \psi(v)$. Therefore, $f(p)$ lies in some open simplex of $L$ for which $\psi(v)$ is a vertex, which does the trick.

**36. Corollary.** If $\psi : K \to L$ is a simplicial approximation to $f : K \to L$ then $d_s(f, \psi) \leq$ mesh $L$.

**37. Lemma.** If $f : K \to L$ is a simplicial map and $\psi : K \to L$ is a simplicial approximation of $f$ then $\psi = f$.

**Proof.** $\psi | K(0) = f | K(0)$.

**38. Theorem.** Let $\psi$ be a simplicial approximation to $f : K \to L$. If $K'$ is a subcomplex of $K$ such that $f | K'$ is a simplicial map, then there exists a homotopy $h : f - \psi$ (fix $K'$) (*i.e.*, $h(x, t) = f(x) = \psi(x)$, for all $x \in K'$).

**Proof.** Define $h : K \times I \to L$ by $h(x, t) = t\psi(x) + (1 - t)f(x)$. By Lemma 35, $h$ is a well-defined map into $L$. Clearly, $h$ is a homotopy between $f$ and $\psi$. Also, $h$ is *stationary* on $K'$ because of Lemma 37.

**39. Lemma.** Let $f : K \to L$ be a continuous map and $\psi : K(0) \to L(0)$ be a vertex map. Then $\psi$ can be extended to a simplicial approximation to $f$ if and only if $f(St\ v) \subset St\ \psi(v)$, for each $v \in K(0)$.

**Proof.** Since the *only if* part is obvious, let us prove the *if* part. We must show that $\psi$ satisfies Definition 33(b): Pick $\langle v_0, \ldots, v_n \rangle \subset K$. Then, by Lemma 20(d) and the hypothesis, $\emptyset \neq f(\langle v_0, \ldots, v_n \rangle) \subset f(\bigcap_{i=0}^{n} St\ v_i) \subset \bigcap_{i=0}^{n} f(St\ v_i) \subset \bigcap_{i=0}^{n} St\ \psi(v_i)$. Therefore, by Lemma 20(d), $\langle \psi(v_0), \ldots, \psi(v_n) \rangle^- \subset L$.

**40. Theorem.** Let $K$ be a finite polytope, $L$ any polytope and $f : K \to L$ a continuous map. Then there exists a subdivision $K^*$ of $K$ (not necessarily a barycentric subdivision) and a simplicial map $\psi : K^* \to L$ such that $\psi$ is a simplicial approximation to $f$.

**Proof.** By Lemma 20(a) and the continuity of $f$, $\mathcal{V} = \{f^{-1}(St\ v) | v \in L(0)\}$ is an open cover of $K$. Since $K$ is a compact metric space (see Corollary 22) $\mathcal{V}$ has a Lebesgue number $\delta > 0$ (see ex. 3.13). Choose $K^*$ so that mesh $K^* < \delta/2$ (see Theorem 32). Then diam $\sigma \leq \delta/2$, for each open simplex $\sigma \subset K^*$; hence, $St\, w \subset B(w, \delta)$, for each $w \in K^*(0)$. Therefore, for each $w \in K^*(0)$, there exists $v(w) \in L(0)$ such that $St\, w \subset B(w, \delta) \subset f^{-1}(St\, v(w))$. Define $\psi : K^*(0) \to L(0)$ by $\psi(w) = v(w)$. Then $f(St\, w) \subset St\, \psi(w)$, for each $w \in K^*(0)$, and we can therefore extend $\psi$ to a simplicial approximation $\psi$ of $f$, by Lemma 39, which completes the proof.

**41.   Corollary.** Let $f : K \to L$ be continuous with $K$ finite and $L$ a finite dimensional polytope. Then, for each $\varepsilon > 0$, there exists (barycentric) subdivisions $K^{(n)}$ of $K$ and $L^{(m)}$ of $L$ and a simplicial approximation $\psi : K^{(n)} \to L^{(m)}$ to $f$ such that $d_s(f, \psi) < \varepsilon$.

**Proof.** By Theorem 32, pick $L^{(m)}$ such that mesh $L^{(m)} < \varepsilon$. Then $f : K \to L^{(m)}$ is continuous, by Lemma 26. Therefore, by Theorem 40, there exists $K^{(n)}$ and a simplicial approximation $\psi : K^{(n)} \to L^{(m)}$ to $f$. By Corollary 36, $d_s(f, \psi) \leq$ mesh $L^{(m)} < \varepsilon$.

## 8.6    The Fundamental Group and Polytopes

Theorems 38 and 40 shows that one may study the homotopy groups of a polytope by considering only simplicial maps and their homotopy properties. This simple observation enables one to effectively compute the fundamental homotopy group of many spaces.

**42. Definition.** Let $K$ and $L$ be polytopes and $\psi, \mu : K \to L$ be simplicial maps. $\psi$ and $\mu$ are *contiguous* if, for each $\langle v_0, \dots, v_k \rangle \subset K$,

$$\langle \psi(v_0), \dots, \psi(v_k), \mu(v_0), \dots, \mu(v_k) \rangle \subset L.$$

The maps $\psi$ and $\mu$ are *contiguous equivalent* (*i.e.*, $\psi \overset{c}{\sim} \mu$ if there exists a finite sequence $\psi_0, \dots, \psi_k : K \to L$ of simplicial maps such that $\psi = \psi_0$, $\mu = \psi_k$ and $\psi_{i-1}$ is contiguous to $\psi_i$, for $i = 1, \dots, k$.

**43. Lemma.** Let $K$ and $L$ be polytopes and $f : K \to L$ a continuous map. If $\psi$, $\mu : K \to L$ are simplicial approximations of $f$ then $\psi$ and $\mu$ are contiguous.

**Proof.** This is an easy consequence of Lemma 20(d).

**44.   Lemma.** Suppose $\psi, \mu : K \to L$ are contiguous simplicial maps and let $A = \{x \in K | \psi(x) = \mu(x)\}$. Then $\psi - \mu$ (fix $A$).

**Proof.** Note that, for each $p \in K$, $\psi(p)$ and $\mu(p)$ lie in a common simplex of $L$.

Define $h : K \times I \to L$ by

$$h(x, t) = (1 - t)\psi(x) + t\mu(x).$$

It is easily seen that $h$ is the desired homotopy.

**45. Corollary.** Contiguous equivalent simplicial maps are homotopic.

**46. Theorem.** Let $K$ be a finite polytope and $L$ any polytope. Let $f_0, f_1 : K \to L$ be continuous maps and $A = \{x \in K | f_0(x) = f_1(x)\}$. Suppose that there exists a homotopy $h : f_0 - f_1$ (fix $A$). Then, for some $n$, there exist simplicial maps $\gamma_0, \gamma_1 : K^{(n)} \to L$ such that

(a) $\gamma_j$ is a simplicial approximation of $f_j$, for $j = 0, 1$,

(b) $\gamma_0 \overset{c}{\sim} \gamma_1$.

**Proof.** By Lemma 20(a) and continuity of $h$, $\{h^{-1}(St\ w) | w \in L(0)\}$ is an open cover of $K \times I$. Since $K \times I$ is a compact metric space there exists $\delta > 0$ such that each $B(x, \delta) \subset h^{-1}(St\ w)$, for some $w \in L(0)$. Choose barycentric subdivisions $K^{(n)}$ of $K$ and $I^{(k)}$ of $I$ fine enough that $St\ v \times \left[j - \frac{1}{2^k}, j + \frac{1}{2^k}\right]$ is contained in a ball of radius $\delta$ and, therefore, it is contained in some $h^{-1}(St\ w)$, (note that the vertices of $I^{(k)}$ are $0, \frac{1}{2^k}, \dots, \frac{2^k - 1}{2^k}, 1$). Since $K^{(n)} \times I^{(k)}$ is clearly homeomorphic to a $CW$-polytope with vertices $\left(v, \frac{1}{2^k}\right)$, for $v \in K^{(n)}(0)$ and $i = 0, \dots, 2^k$, and

$$St\left(v, \frac{i}{2^k}\right) \subset St\ v \times \left[\frac{i-1}{2^k}, \frac{i+1}{2^k}\right] \subset \text{some } h^{-1}(St\ w).$$

there exists, by Lemma 39, a simplicial approximation $\mu : K^{(n)} \times I^{(k)} \to L$ of $h$. Note that, by Lemma 39,

$$St\ v \times \left[\frac{i-1}{2^k}\frac{i+1}{2^k}\right] \subset h^{-1}\left(St\ \mu\left(v, \frac{i}{2^k}\right)\right).$$

Let $\psi_i = \mu | K^{(n)} \times \{\frac{i}{2^k}\}$, for $i = 0, \dots, 2^k$. Then, letting $\phi_0 = \psi_0$, $\phi_1 = \mu_2$, $\phi_j$ is a simplicial approximation of $f_j$, for $j = 0, 1$. Furthermore, $\phi_0 \overset{c}{\sim} \phi_1$ because $\psi_i$ is contiguous to $\psi_{i+1}$, for $i = 0, \dots, 2^k - 1$: Pick any simplex $\langle v_0, \dots, v_m \rangle \in K^{(n)}$. Then

$$\bigcap_{j=0}^{m} St\ \mu\left(v_j, \frac{i}{2^k}\right) \cap \bigcap_{j=0}^{m} St\ \mu\left(v_j, \frac{i+1}{2^k}\right)$$

$$\supset \bigcap_{j=0}^{m} h\left(St\ v_j \times \left[\frac{i-1}{2^k}, \frac{i+1}{2^k}\right]\right)$$

$$\cap \bigcap_{j=0}^{m} h\left(St\ v_j \times \left[\frac{i}{2^k}, \frac{i+2}{2^k}\right]\right)$$

$$\cap h\left(St\ v_j \times \left[\frac{i}{2^k}, \frac{i+1}{2^k}\right]\right) \neq \emptyset,$$

and this shows that $\langle \psi_i(v_0), \ldots, \psi_i(v_m), \psi_{i+1}(v_0), \ldots, \psi_{i+1}(v_m) \rangle \subset L$; therefore $\psi_i$ is contiguous to $\psi_{i+1}$.

**47. Definition.** Let $K$ be a polytope.

(a) An ordered pair $|v_1 v_2|$ of vertices of $K$, such that $\langle v_1, v_2 \rangle \subset K$, is called an *edge* of $K$ with origin $v_1$ and end $v_2$. If $e = |v_1 v_2|$ then $e^{-1} = |v_2 v_1|$.

(b) A *route* in $K$ is a finite sequence $\omega = e_1 e_2 \ldots e_n$ of edges of $K$ such that the origin of $e_{i+1}$ is the end of $e_i$, for $i = 1, \ldots, n - l$. The origin of $e_1$ is the origin of $\omega$ and the end of $e_n$ is the end of $\omega$.

(c) Given two routes $\omega = e_1 e_2 \ldots e_k$ and $\sigma = d_1 d_2 \ldots d_n$, with the end of $\omega$ equal to the origin of $\sigma$, we define the *product $\omega\sigma$* by
$$\omega\sigma = e_1 e_2 \ldots e_k d_1 d_2 \ldots d_n.$$

(d) The *inverse* of a route $\omega = e_1 \ldots e_k$ is $\omega^{-1} = e_k^{-1} \ldots e_1^{-1}$.

(e) For any three vertices $v_1, v_2, v_3$ of a simplex of $K$, we say that $|v_1 v_2||v_2 v_3|$ is *edge equivalent* to $|v_1 v_3|$. Two routes $\omega$ and $\sigma$ are *edge equivalent* (*i.e.,* $\omega \overset{E}{\sim} \sigma$) if $\sigma$ can be obtained from $\omega$ by a sequence of elementary edge equivalences.

**48. Theorem.** Let $K$ be a polytope and $v_0$ a vertex of $K$. Let $E(K, v_0)$ be the set of edge equivalence classes of routes of $K$ with origin and end at $v_0$. Then $E(K, v_0)$ is a group, with identity $|v_0 v_0|$, under the operation of multiplication and inverse defined above. ($E(K, v_0)$ is called the *edge path group* of $(K, v_0)$.)

**Proof.** Straightforward.

By its very definition, the edge path group of $(K, v_0)$ depends only on the *simplexes* of $K$ and not on the topology of $K$.

**49. Theorem.** Let $K$ be a polytope and $v_0$ a vertex of $K$. Then $E(K, v_0)$ and $\prod_1(K, v_0)$ are isomorphic groups.

**Proof.** We define homomorphisms $h : E(K, v_0) \to \prod_1(K, v_0)$ and $g : \prod_1(K, v_0) \to E(K, v_0)$ such that $g \circ h = 1_{E(K,v_0)}$ and $h \circ g = 1_{\prod_1(K,v_0)}$. This will show that both $g$ and $h$ are isomorphisms.

The construction of $h$: Let $[\omega] \in E(K, v_0)$. Then $\omega = |v_0 v_1||v_1 v_2||v_2 v_3| \cdots |v_{n-1} v_n|$, with $v_n = v_0$, for some $\{v_0, \ldots, v_n\} \subset K(0)$. Regard $I$ as a complex with vertices $\{0, \frac{1}{n}, \ldots, \frac{n-1}{n}, 1\}$ and consider the vertex map $\bar{\omega} : I(0) \to K(0)$ defined by $\bar{\omega}(j/n) = v_j$, for $j = 0, \ldots, n$. Since $\omega$ is a route, extend $\bar{\omega}$ to a simplicial map $\bar{\omega} : I \to K$ (we use the same symbol!). Let $h([\omega]) = [\bar{\omega}]$. Since $\omega \overset{E}{\sim} \sigma$ implies that $\bar{\omega} \underset{\tilde{v}_0}{\sim} \bar{\sigma}$ (by Corollary 45), we get that $h$ is well-defined. It is also easily seen that $h$ is a homomorphism (If $\omega = e_1 \ldots e_k$ and $\sigma = d_1 \ldots d_n$ are routes with origin and end $v_0$, we define a homotopy between $\overline{\omega\sigma}$ and $\bar{\omega} * \bar{\sigma}$ by changing the travelling time of the $e_i$ and $d_i$ from $\frac{1}{k+n}$ to $\frac{1}{2k}$ and $\frac{1}{2n}$, respectively, in the homotopy square).

The construction of $g$: Pick $[\alpha] \in \prod_1(K, v_0)$ and some simplicial approximation $\psi_\alpha : I^{(n)} \to K$ of $\alpha$, for some subdivision $\{0, \frac{1}{n}, \ldots, \frac{n-1}{n}, 1\}$ of $I$. Then $\psi_\alpha \underset{\tilde{v}_0}{\sim} \alpha$ and $\psi_\alpha$ defines an edge path $\psi'_\alpha$ starting and ending at $v_0$. Let $g([\alpha]) = [\psi'_\alpha]$. Note that, by Theorem 46, $\alpha \underset{\tilde{v}_0}{\sim} \sigma$ implies that $\psi_\alpha \overset{c}{\sim} \psi_\alpha$, which in turn implies that $\psi'_\alpha \overset{\varepsilon}{\sim} \psi'_\alpha$. Therefore, $g$ is well-defined. Even though $g$ is a homeomorphism, we will not need this information. Clearly,

$$g \circ h = 1_{E(K, v_0)} \text{ and } h \circ g = 1_{\prod_1(K, v_0)}$$

and, therefore, one easily sees that $h$ is one-to-one and onto. Consequently $E(K, v_0) \approx \prod_1(K, v_0)$.

**50. Corollary.** Let $K$ be a polytope and $v_0 \in K(0)$. Let $i : K(2) \to K$ be the injection map. Then $i$ induces an isomorphism $i_* : E(K(2), v_0) \to E(K, v_0)$. Therefore $\prod_1(K(2), v_0) \approx \prod_1(K, v_0)$.

## 8.7 Graphs and Trees

**51. Definition.** A *graph* is a polytope of dimension less than 2. A *tree* $T$ is an arcwise connected graph such that $T - s$ is not connected, for each *open* 1-simplex $s \subset T$. An *end* of a graph is a vertex which is a vertex of at most one 1-simplex.

**52. Lemma.** Let $K$ be a connected polytope. Then $K$ contains a maximal (with respect to inclusion) tree and any maximal tree contains all vertices of $K$.

**Proof.** Partially order the collection of all trees contained in $K$ by inclusion. Pick a nest $\{T_\alpha\}$ of trees and let $T = \bigcup_\alpha T_\alpha$. Clearly $T$ is arcwise connected. Let $\langle v_0, v_1 \rangle$ be any *open* 1-simplex contained in $T$. If $T - \langle v_0, v_1 \rangle$ is connected, then there exists an edge path $\omega$ starting at $v_0$ and ending at $v_1$ which does not use the edge $|v_0 v_1|$. Then $|v_1 v_0|\omega$ is a *closed circuit* which is contained in some $T_\alpha$ (therefore $T_\alpha - \langle v_0, v_1 \rangle$ is connected), a contradiction. This shows that $T$ is a tree. By Zorn's Lemma, we get that $K$ contains a maximal tree.

Because of the connectedness of $K$, one easily sees that any maximal tree in $K$ contains all vertices of $K$.

**53. Definition.** Let $T$ be a maximal tree of the connected polytope $K$. Let $E(K - T)$ be the group generated by the edges $|vu|$ of $K$ with the relations.

(a) If $|vu|$ is an edge of $T$ then $|vu| = 1$.
(b) If $v_0$, $v_1$, $v_2$ are vertices of a simplex of $K$ then $|v_0 v_1||v_1 v_2| = |v_0 v_2|$.

**54. Theorem.** $E(K, v_0) \approx E(K - T)$.

**Proof.** Since $T$ is connected and $T \supset K(0)$, for each $v \in K(0)$, there exists an edge path $\Gamma_v$, with origin $v_0$ and end $v$, which is contained in $T$. Note that, for each edge $|vu|$ of $K$, the edge path $\Gamma_v |vu| \Gamma_u^{-1}$ starts and ends at $v_0$.

Define $h : E(K, v_0) \to E(K - T)$ by

$$h([|v_0v_1||v_1v_2|\cdots|v_{k-1}v_k|]) = |v_0v_1||v_1v_2|\cdots|v_{k-1}v_k|.$$

Note that $h$ is well-defined, because of Definition 53(b) and the definition of edge equivalence. Clearly, $h$ is a homomorphism (note that $h([|v_0v_0|]) = |v_0v_0| = 1$ because of Definition 53(a)). Let us also observe that

(1) $h$ is onto: Let $|v_1v_2||v_3v_4|\cdots|v_nv_{n+l}| \in E(K - T)$ and let $w$ be the equivalence class of

$$\Gamma_{v_1}|v_1v_2|\Gamma_{v_2}^{-1}\Gamma_{v_3}|v_3v_4|\Gamma_{v_4}^{-1}\cdots\Gamma_{v_n}|v_nv_{n+1}|\Gamma_{v_{n+1}}^{-1} \text{ in } E(K, v_0).$$

Then $h(|w|) = |v_1v_2||v_3v_4|\cdots|v_nv_{n+1}|$ because of Definition 53(a) and the fact that $\Gamma_{v_i} \subset T$, for $i = 1, \ldots, n + 1$. This shows that $h$ is onto.

(2) $h$ is one-to-one: let $[w] \in E(K, v_0)$ with $\omega = |v_0v_1|\cdots|v_{n-1}v_0|$, and suppose that $h([w]) = 1$. If $|v_0v_1|, \ldots, |v_{n-1}v_0| \in T$ then one immediately gets that $\omega \overset{E}{\sim} |v_0v_0|$ (*i.e.*, $[w] = [|v_0v_0|]$). If not all $|v_0v_1|, \ldots, |v_{n-1}v_0| \in T$ then, from Definition 53(b) and the fact that $|v_0v_1||v_1v_2|\ldots|v_{n-1}v_0| = 1$, we get that $|v_0v_1||v_1v_2|\ldots|v_{n-1}v_0| \overset{E}{\sim} |v_0v_0|$. Therefore, we have proved that $\text{Ker}\, h = [|v_0, v_0|]$ (*i.e.*, $h$ is one-to-one), which completes the proof.

**55. Corollary.** If $K$ is a connected graph then $E(K, v_0)$ is a free group. If $T$ is a maximal tree in $K$ then the generators of $E(K, v_0)$ are in one-to-one correspondence with the 1-simplexes of $K - T$.

**Proof.** It suffices to show that $E(K - T)$ is a free group generated by the 1-simplexes of $K - T$. Because of Definition 53(a), $E(K - T)$ is generated by the 1-simplexes of $K - T$ (note that if $e = |vu|$ is an edge in $K - T$ then its inverse in $E(K - T)$ is $e^{-1} = |uv|$). Also, there are no relations of the form described in Definition 53(b) between any two edges $|vu|$ and $|uw|$ of $K - T$ with $v$, $u$ and $w$ vertices of some simplex of $K$ (since $K$ is one-dimensional, either $u = v$ or $v = w$ or $u = w$; if $u = v$ or $v = w$, then $|uv||vw| = |uw|$ or $|uv||vw| = |uv|$; if $u = w$, then $|uv||vw| = |uv||vu| = e^{-1} = 1$). Therefore, $E(K - T)$ is freely generated by the 1-simplexes of $K - T$, which completes the proof.

The following result is an immediate consequence of Corollary 55.

**56. Theorem.** The following is true:

(a) $\prod_1(S^1) \approx Z$ (the group of integers)—see ex. 11.

(b) $\prod_1$ (Figure Eight) $\approx F_2$ (where $F_n$ denotes a free group with $n$ generators).

**Chapter 8.  Exercises**

1. Let $\Omega(X)$ be $X^I$ with the co topology. Prove that

(a) If $\alpha$, $\beta \in \Omega(X)$ and $h : \alpha \sim \beta$, then the map $\psi : I \to \Omega(X)$, defined by $(\psi(s))(t) = h(t, s)$, is an arc in $\Omega(X)$ from $\alpha$ to $\beta$.

(b) If $\gamma : I \to \Omega(X)$ is an arc from $\alpha$ to $\beta$, then the map $h : I \times I \to X$, defined by $h(s,t) = (\gamma(t))(s)$, is a homotopy from $\alpha$ to $\beta$.

(c) Parts (a) and (b) remain valid for $\Omega(X,b)$ and $\tilde{b}$.

2. Let $X$ be the disjoint union of a copy of $S^1$ and a copy of $B^2$. Let $p \in S^1$ and $q \in B^2$. Show that $\prod_1(X,p) = Z$ and $\prod_1(X,q) = 1$.

3. Check that $h_2 : I \times I \to Y$ defined by

$$h_2(s,t) = \begin{cases} \sigma^{-1}(2s), & s \le \frac{1-t}{2}, \\ h_1\left(\frac{4s+2t-2}{3t+1}, t\right), & \frac{1-t}{2} \le s \le \frac{t+3}{4}, \\ \sigma(4s-3), & s \ge \frac{t+3}{4}, \end{cases}$$

is a homotopy satisfying the requirements in the proof of Theorem 11(b).

4. Let $X = \bigcup_\alpha \in_\Lambda X_\alpha$ and $\tau_\alpha$ be a topology for $X_\alpha$, for each $\alpha \in \Lambda$. Let $\tau$ be the family of all $U \subset X$ such that $U \cap X_\alpha \in \tau_\alpha$, for each $\alpha \in \Lambda$.

(a) Show that $\tau$ is a topology for $X$.

(b) Show that, in general, $(X_\alpha, \tau_\alpha)$ may not be a subspace of $(X, \tau)$. (Hint: Let $X = X_1 \cup X_2$ and let $\tau_1$ and $\tau_2$ be topologies on $X_1$ and $X_2$, respectively, such that $\tau_1|(X_1 \cap X_2) \ne \tau_2|(X_1 \cap X_2)$.)

5. Take a simplex $\langle x_0, \ldots, x_n \rangle^-$ and let $B^n = \{(y_0, \ldots, y_{n-1}) \in E^n | y_0^2 + \cdots + y_{n-1}^2 \le 1\}$. Define a map $h : B^n \to \langle x_0, \ldots, x_n \rangle^-$ by $h(\bar{y}) = \sum_{i=0}^{n-1} y_i^2 x_i + (1 - |\bar{y}|^2)x_n$. Prove that

(1) $h$ is onto (solve a system of $n$ equations with $n+1$ unknowns).

(2) $h$ is not $1-1$.

(3) $h(S^{n-1}) = \langle x_0, \ldots, x_{n-1} \rangle^-$.

(4) Is $h|S^{n-1}$ a $1-1$ function?

6. Let $K$ be a simplicial complex with the weak topology and let $f : K \to Y$ be a function to a topological space $Y$. Show that, if $f|s$ is continuous, for each closed simplex $s$ of $K$, then $f$ is continuous. Show that the *identity* functions $K \to K^{(1)}$ and $K^{(1)} \to K$ are continuous.

7. Show that, by Theorem 49,

(a) $\prod_1(\text{Torus}) \approx Z \times Z$

(b) $\prod_1(\text{Möbius Band}) \approx Z$

(c) $\prod_1(\text{Klein Bottle}) \approx$ A group generated by two elements $a$, $b$ with $a^2b^2 = 1$.

8. Show that the following triangulation is not a triangulation of a torus; indeed, by eliminating *repeated* triangles, it is a triangulation of a cylinder. (Hint: Note that the triangles labeled A and B have the same vertices.)

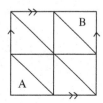

9. Show that the following is a triangulation of a torus; all triangles are *distinct*.

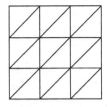

10. In $E^n$, let $x_0 = (1, 0, \ldots, 0) = e_1$, $x_1 = (0, 1, 0, \ldots, 0) = e_2, \ldots x_{n-1} = e_n$ and $x_n = \left(-\frac{1}{\sqrt{n}}, -\frac{1}{\sqrt{n}}, \ldots, -\frac{1}{\sqrt{n}}\right)$. Also for each $\bar{y} = (y_0, \ldots, y_{n-1}) \in S^{n-1} \subset B^n$, let

$$t_{\bar{y}} = \sup\{s \mid s \in E^1 \text{ and } s\bar{y} \in \langle x_0, \ldots, x_n \rangle^-\}.$$

Define a function $h : B^n \to E^n$ by

$$h(s\bar{y}) = st_{\bar{y}}\bar{y},$$

for each $\bar{y} \in S^{n-1}$ and $0 \leq s \leq 1$. Prove that

(a) $\{x_0, x_1, \ldots, x_n\}$ is a convex-independent subset of $B^n$.

(b) $\langle x_0, \ldots, x_n \rangle^- \subset B^n$ and $\bar{0} \in \langle x_0, \ldots, x_n \rangle$.

(c) $h(B^n) = \langle x_0, \ldots, x_n \rangle$.

(d) $h$ is $1 - 1$.

(e) $h : B^n \to \langle x_0, \ldots, x_n \rangle^-$ is a homeomorphism.

(f) $h|S^{n-1} : S^{n-1} \to \partial\langle x_0, \ldots, x_n \rangle^-$, where $\partial\langle x_0, \ldots, x_n \rangle^-$ is the polytope of all proper faces of $\langle x_0, \ldots, x_n \rangle^-$.

11. Show that $\prod_1(S^1) \approx Z$. (Hint: Use Theorem 54 and part (f) of the preceding exercise.)

12. Show that $\prod_1(S^n) = 1$ for $n \geq 2$. (Hint: Use ex. 10(f); start with $n = 2$, using Theorem 54, and use induction.)

# Appendix A

# Some Inequalities

We limit ourselves to those inequalities crucial to various proofs in topology.

**Cauchy-Schwartz** inequality: If $a_1, \ldots, a_n$ and $b_1, \ldots, b_n$ are real numbers then

$$\left( \sum_k a_k b_k \right)^2 \leq \left( \sum_k a_k^2 \right) \left( \sum_k b_k^2 \right).$$

**Proof.** Clearly $\sum_k (a_k x + b_k)^2 \geq 0$, for every $x \in E^1$. Therefore, $0 \leq \sum_k (a_k x + b_k)^2 = Ax^2 + 2Bx + C$, with $A = \sum_k a_k^2$, $B = \sum_k a_k b_k$, $C = \sum_k b_k^2$. If $A = 0$, clearly $B^2 \leq AC$. If $A > 0$, let $x = -\frac{B}{A}$. Then $0 \leq Ax^2 + 2Bx + C = -\frac{B^2}{A} + C$, which implies that $B^2 - AC \leq 0$. This completes the proof.

**Minkowski** inequality. If $a_1, \ldots, a_n$ and $b_1, \ldots, b_n$ are real numbers, then

$$\left( \sum_k (a_k + b_k)^2 \right)^{1/2} \leq \left( \sum_k a_k^2 \right)^{1/2} + \left( \sum_k b_k^2 \right)^{1/2}.$$

**Proof.** Note that $\left( \sum_k (a_k + b_k)^2 \right)^{1/2} = \left( \sum_k (a_k^2 + 2a_k b_k + b_k^2) \right)^{1/2} = \left( \sum_k a_k^2 + 2 \left( \left( \sum_k a_k b_k \right)^2 \right)^{1/2} + \sum_k b_k^2 \right)^{1/2} \leq \left( \sum_k a_k^2 + 2 \left( \sum_k a_k^2 \right)^{1/2} \left( \sum_k b_k^2 \right)^{1/2} + \sum_k b_k^2 \right)^{1/2}$ (by Cauchy-Schwartz inequality) $= \left( \left[ \left( \sum_k a_k^2 \right)^{1/2} + \left( \sum_k b_k^2 \right)^{1/2} \right]^2 \right)^{1/2} = \left( \sum_k a_k^2 \right)^{1/2} + \left( \sum_k b_k^2 \right)^{1/2}$. This completes the proof.

# Appendix B

# Binomial Equalities

We limit ourselves to the consequences of the Binomial Expansion which we need. The techniques clearly indicate that the *binomial identities* are endless.

**Proposition 1.** (Binomial Expansion). Given any real numbers $x$ and $y$ and $n \in \mathbf{N}$,

$$(x + y)^n = \sum_{j=0}^{n} \binom{n}{j} x^j y^{n-j},$$

with $\binom{n}{j} = \frac{n!}{j!(n-j)!}$.

**Proof.** Elementary induction.

For $n \in \mathbf{N}$, define a function $B_n : E^2 \to E^1$, by letting $B_n(x, y) = (x + y)^n$. Then, we get

**Proposition 2.** $x(x + y)^{n-1} = \sum_{j=0}^{n} \frac{j}{n} \binom{n}{j} x^j y^{n-j}$, for each $(x, y) \in E^2$ and $n \in \mathbf{N}$.

**Proof.** Differentiating $B_n$ with respect to $x$ and using Proposition 1, we get that

$$\frac{\partial B_n}{\partial x} = n(x + y)^{n-1} = \frac{\partial}{\partial x} \left( \sum_{j=0}^{n} \binom{n}{j} x^j y^{n-1} \right)$$

$$= \sum_{j=0}^{n} \binom{n}{j} j x^{j-1} y^{n-j},$$

from which the result immediately follows.

**Proposition 3.** $\left(1 - \frac{1}{n}\right) x^2 (x + y)^{n-2} = \sum_{j=0}^{n} \left(\frac{j^2}{n^2} - \frac{j}{n^2}\right) \binom{n}{j} x^j y^{n-j}$, for every $(x, y) \in E^2$ and $n \in \mathbf{N}$.

**Proof.** Essentially the same as the proof of Proposition 2, except that we compute $\partial^2 B_n / \partial x^2$.

**Proposition 4.** $\sum_{j=0}^{n} \binom{n}{j} x^j (1-x)^{n-j} = 1$, for each $x \in E^1$.

**Proof.** Apply Proposition 1 to $1 = (x + (1-x))^n$.

**Proposition 5.** $x = \sum_{j=0}^{n} \frac{j}{n} \binom{n}{j} x^j (1-x)^{n-j}$, for each $x \in E^1$.

**Proof.** Apply Proposition 2 to $x = x(x + (1-x))^n$.

**Proposition 6.** $\left(1 - \frac{1}{n}\right) x^2 + \frac{1}{n} x = \sum_{j=0}^{n} \frac{j^2}{n^2} \binom{n}{j} x^j (1-x)^{n-j}$, for each $x \in E^1$.

**Proof.** Apply Proposition 3 to $\left(1 - \frac{1}{n}\right) x^2 (x + (1-x))^{n-2}$.

**Proposition 7.** $\sum_{j=0}^{n} \left(x - \frac{j}{n}\right)^2 \binom{n}{j} x^j (1-x)^{n-j} \leq \frac{1}{4n}$, for each $x \in E^1$.

**Proof.** Note that

$$\sum_{j=0}^{n} \left(x - \frac{j}{n}\right)^2 \binom{n}{j} x^j (1-x)^{n-j} = \sum_{j=0}^{n} \left(x^2 - \frac{2xj}{n} + \frac{j^2}{n^2}\right) \binom{n}{j} x^j (1-x)^{n-j}$$

$$= x^2 \sum_{j=0}^{n} \binom{n}{j} x^j (1-x)^{n-j}$$

$$+ \sum_{j=0}^{n} \frac{j^2}{n^2} \binom{n}{j} x^j (1-x)^{n-j}$$

$$- 2x \sum_{j=0}^{n} \frac{j}{n} \binom{n}{j} x^j (1-x)^{n-j}$$

$$= x^2 + \left[\left(1 - \frac{1}{n}\right) x^2 + \frac{1}{n} x\right] - 2x^2$$

$$= \frac{x(1-x)}{n} \leq \frac{1}{4n},$$

because $\sup_{x \in E^1} x(1-x) = \frac{1}{4}$ (the function $f(x) = x(1-x)$ has a maximum value at $x = 1/2$).

# List of Symbols

# Index

Printed in the United States
by Baker & Taylor Publisher Services

Printed in the United States
by Baker & Taylor Publisher Services